# Sodium Minerals

# Sodium Minerals

## Unearthing The Geological Treasures

Mack Rafeal

Bliss Enterprise

# CONTENTS

# INDEX

# INTRODUCTION

In the complicated embroidery of Earth's land piece, sodium minerals stand as covered up treasures, frequently ignored in the excellent account of mineralogy. These honest mixtures, wealthy in sodium, play quietly played crucial parts in mankind's set of experiences, industry, and the regular world. As we set out on this investigation, the purpose is to dig profound into the geographical domains that harbor these sodium-rich stores, disclosing the privileged insights of their development, mining, applications, and the significant effect they have on our worldwide scene.

## 1.1 The Meaning of Sodium Minerals

To the undeveloped eye, sodium might appear to be a typical component, promptly connected with table salt and ordinary culinary applications. Nonetheless, past this natural setting lies a huge swath of sodium minerals, each with its particular properties and utilities. Sodium minerals, enveloping mixtures like halite, soft drink debris, and sodium feldspar, act as fundamental unrefined components in different businesses, going from assembling and innovation to energy creation.

Understanding the meaning of sodium minerals requires a more extensive focal point, one that catches their verifiable significance and contemporary pertinence. For quite a long time, these minerals have been woven into the texture of human life, molding social practices, shipping lanes, and financial scenes. Uncovering the topographical fortunes of sodium minerals is, in this way, an undertaking not simply logical yet in addition verifiable, social, and financial.

## 1.2 The Verifiable Embroidery of Sodium Minerals

The foundations of sodium mineral utilization follow back to old civilizations, where the extraction and exchange of salt held colossal vital and monetary worth. Early human settlements close to saline sources were many times flourishing centers of exchange and social trade. The old Romans, perceiving the additive characteristics of salt, utilized it widely in food protection — a training that impacted the improvement of shipping lanes and whole economies.

As hundreds of years advanced, the significance of sodium minerals extended past simple food. The Modern Unrest denoted an essential point, catapulting the interest for sodium builds to exceptional levels. Salt creation turned into a key part of industrialization, supporting the assembling of glass, cleanser, materials, and a bunch of substance items. This verifiable setting makes way for our investigation, featuring the getting through effect of sodium minerals on the course of human civilization.

### 1.3 Reason and Extent of the Investigation

The reason for this far reaching investigation is to unwind the geographical complexities encompassing sodium minerals, revealing insight into their development, extraction, applications, monetary importance, and ecological ramifications. By stripping back the layers of the World's hull, we expect to uncover the land loves that support current modern cycles and innovative headways.

The extent of this investigation reaches out past the limits of customary mineralogical studies. We look to overcome any barrier between land sciences and the more extensive range of human undertakings affected by sodium minerals. Through this interdisciplinary methodology, we desire to give a comprehensive comprehension of sodium minerals, their part in molding our reality, and the potential they hold for future reasonable turn of events.

### 1.4 Construction of the Investigation

This investigation is organized to unfurl like the layers of sedimentary stone, every section uncovering an alternate feature of sodium minerals. Starting with a top to bottom gander at the science and qualities of sodium minerals, we will travel through their land development, the strategies utilized in mining and extraction, their different applications across businesses, and the monetary ramifications on a worldwide scale.

Parts will investigate the fragile harmony between saddling the financial advantages of sodium minerals and the basic of maintainable practices. We will analyze the natural effect of sodium mineral extraction and handling, alongside procedures for alleviating these impacts. As we peer into the future, the investigation will finish up by thinking about arising innovations, developments, and the likely job of sodium minerals in molding a maintainable and green future.

### 1.5 Exploring the Geographical Scene

This excursion into the geographical fortunes of sodium minerals welcomes perusers to explore the complex scene of Earth's structure. Whether you are a geologist, an industry proficient, a history specialist, or a natural promoter, this investigation means to give a complete and open manual for sodium minerals. As we leave on this odyssey, we welcome you to uncover the secret stories inside the Earth, where sodium minerals unobtrusively hold the keys to grasping our past, molding our present, and fashioning a maintainable future.

### 1. Definition and significance of sodium minerals

In the huge scene of Earth's geographical fortunes, sodium minerals arise as overlooked yet truly great individuals, discreetly applying their effect on a large number of human exercises and ventures. To see the value in their importance completely, it is basic to dig into the definition and innate significance of sodium minerals.

This investigation will disentangle the unmistakable attributes that characterize sodium minerals and shed light on the significant importance they hold in different circles of our lives.

## 1.1 Characterizing Sodium Minerals

At its basic center, sodium is a responsive salt metal with nuclear number 11, represented as Na on the occasional table. Nonetheless, the genuine wonder lies in the bunch intensifies that sodium structures with different components, aggregately known as sodium minerals. These minerals include a different reach, from the notable halite (normal table salt) to soft drink debris, sodium feldspar, and then some. The principal quality is the presence of sodium as an essential constituent, granting novel properties and applications to these minerals.

## 1.2 The Synthetic Embroidery of Sodium Minerals

Sodium minerals manifest a rich and mind boggling substance embroidery. The most pervasive among them, halite (NaCl), remains as a demonstration of the key job of sodium in regular day to day existence. Its culinary significance as a flavoring specialist veils the more extensive meaning of sodium minerals in different ventures. Soft drink debris, or sodium carbonate ($Na_2CO_3$), tracks down broad use in the creation of glass, cleansers, and different synthetic compounds. In the mean time, sodium feldspar ($NaAlSi_3O_8$) assumes a crucial part in the ceramics and glass industry.

The variety of sodium minerals emerges not just from the mixes of sodium with different components yet in addition from their particular translucent designs and actual properties. This compound variety converts into an expansive range of uses that length enterprises and add to the headway of innovation.

## 2. The Meaning of Sodium Minerals

## 2.1 Culinary and Additive Importance

While the everyday utilization of table salt might appear to be typical, its part in human nourishment and food safeguarding is principal. Sodium, as halite, has been a basic part of food safeguarding for centuries. The capacity of salt to hinder microbial development has been saddled by different societies since forever ago, forming culinary practices and impacting the course of shipping lanes and economies.

## 2.2 Modern Applications

The modern meaning of sodium minerals is a foundation of current assembling processes. Soft drink debris, a subordinate of sodium carbonate, is a critical part in the creation of glass. The clearness and strength it grants to glass have made it crucial in compositional, auto, and holder glass producing.

Furthermore, sodium compounds are vital in the creation of cleansers, materials, and synthetics, delineating their unavoidable impact on different modern areas.

## 2.3 Innovative Headways

Past the customary domains of industry, sodium minerals add to state of the art innovative progressions. Sodium-particle batteries, for example, address a boondocks where sodium's electrochemical properties are bridled for energy capacity. As the interest for feasible energy arrangements develops, sodium minerals might assume an instrumental part in molding the scene of environmentally friendly power advances.

## 3. Financial Ramifications of Sodium Minerals

## 3.1 Worldwide Market Patterns

The financial meaning of sodium minerals stretches out across borders, with worldwide market patterns mirroring their imperativeness. The interest for sodium minerals is unpredictably connected to the development of modern areas, mechanical advancements, and populace extension. Understanding these market elements gives bits of knowledge into the financial powers driving the extraction, handling, and exchange of sodium minerals on a worldwide scale.

## 3.2 Provincial Effect

Locales supplied with huge sodium mineral stores frequently experience periods of prosperity driven by mining exercises. These stores become geographical highlights as well as financial impetuses, affecting work rates, framework improvement, and provincial thriving. On the other hand, financial slumps in sodium-subordinate businesses can have flowing consequences for neighborhood economies, featuring the fragile harmony between land overflow and monetary dependability.

## 3.3 Difficulties and Open doors

The monetary account of sodium minerals isn't without challenges. Fluctuating business sector costs, ecological worries, and international elements can present obstacles to the supported development of sodium-related ventures. In any case, inside these difficulties lie open doors for advancement, reasonable practices, and the improvement of new business sectors for sodium minerals.

## 4. Ecological Contemplations

## 4.1 Extraction and Handling Effect

Likewise with any extractive industry, the extraction and handling of sodium minerals convey natural ramifications. Mining tasks, especially those including arrangement digging for salts, can influence nearby biological systems and water assets.

Also, energy-serious cycles in the extraction and refinement of sodium minerals add to the carbon impression of sodium-related enterprises.

## 4.2 Feasible Practices and Moderation Procedures

Tending to the ecological worries related with sodium mineral extraction requires a pledge to manageable practices. From limiting the biological effect of mining activities to creating energy-proficient handling techniques, the business can embrace measures to relieve its ecological impression. Reusing and reusing sodium mixtures can additionally add to manageable works on, adjusting sodium enterprises to worldwide endeavors toward natural preservation.

### 4.3 Administrative Systems and Industry Obligation

The ecological effect of sodium minerals requires a hearty administrative system. States and worldwide bodies assume an essential part in characterizing and upholding norms that guide dependable mining and handling rehearses. At the same time, industry partners should bear the obligation of taking on ecologically cognizant methodologies, embracing development, and adding to the worldwide basic of feasible asset the executives.

2. **Historical perspective on sodium mineral usage**

The verifiable woven artwork of sodium minerals is entwined with the advancement of human development, molding social practices, shipping lanes, and financial scenes. From old culinary customs to the modern upheavals of the cutting edge period, the utilization of sodium minerals has been a quiet power, frequently ignored yet significantly effective. This investigation dives into the authentic point of view of sodium mineral utilization, following the strides of human advancements as they found, saddled, and respected these topographical fortunes.

### 1.1 The Beginning of Sodium in Olden times

The starting points of sodium mineral use can be followed back to artifact, where the meaning of salt, specifically, rose above simple culinary purposes. Early human settlements, frequently settled close to saline sources, perceived the additive characteristics of salt, an essential ware in the conservation of food. In old China, salt creation turned into a state syndication, mirroring the essential significance of this mineral in supporting populaces and encouraging exchange.

### 1.2 Salt as Cash and Ware

In different societies, salt expected a double job as both a cash and an item. The word 'compensation' itself finds its foundations in the Latin word 'salarium,' which alluded to the cash distributed to Roman fighters to buy salt.

In bygone eras, salt turned into an important exchange ware, prompting the foundation of salt courses and the ascent of salt towns. The unpredictable associations between salt, riches, and influence highlighted its social and monetary importance.

### 1.3 The Zest of Life and the Time of Investigation

As pioneers set out on journeys of disclosure during the Period of Investigation, salt assumed an essential part as both an additive and a flavoring. Past its pragmatic purposes, salt became emblematic of flourishing and extravagance. It turned into a significant exchange thing districts where it was scant, and salt duties turned into a wellspring of income for some legislatures. The zest courses that associated landmasses likewise worked with the trading of sodium-rich minerals, forming worldwide exchange and social dispersion.

## 2. Industrialization and the Soluble base Unrest

### 2.1 The Modern Insurgency and the Blast of Soluble base Creation

The eighteenth and nineteenth hundreds of years denoted a seismic change in sodium mineral use with the coming of the Modern Unrest. The interest for soluble base, got from sodium minerals, took off as modern cycles thrived. Salt, principally as soft drink debris (sodium carbonate), turned into a key part of different ventures, from cleanser and glass to materials and metallurgy.

### 2.2 Leblanc Interaction and the Introduction of the Compound Business

In 1791, Nicolas Leblanc's development of the Leblanc cycle changed antacid creation. This strategy took into consideration the huge scope union of sodium carbonate from normal salt, making ready for the large scale manufacturing of soluble base. The Leblanc cycle turned into a foundation of the arising compound industry, pushing sodium minerals into the front line of industrialization.

## 3. Sodium Minerals in the Advanced Period

### 3.1 Universal Conflicts and the Essential Significance of Sodium

The twentieth century saw sodium minerals accepting vital significance during both Universal Conflicts. Sodium compounds were pivotal in the development of synthetic compounds, explosives, and engineered materials. The interest for sodium minerals flooded, prompting progressions in extraction advances and the foundation of huge scope mining activities.

### 3.2 From Space Investigation to Regular Innovation

The mid-twentieth century saw sodium minerals entering the domain of room investigation. Sodium-potassium amalgam was utilized in early space missions, showing the flexibility of sodium intensifies in outrageous conditions.

At the same time, sodium's properties tracked down applications in regular advances, from sodium-fume lights to sodium-particle batteries, featuring their flexibility in different fields.

## 4. Social Importance and Imagery

### 4.1 Customs and Imagery in Various Societies

Past their utilitarian jobs, sodium minerals have held social importance and imagery across different social orders. In many societies, salt has been an image of virtue, fellowship, and cordiality. Ceremonies including the sharing of salt, like the antiquated custom of eating with salt, delineate the social significance ascribed to sodium minerals.

### 4.2 Legends and Folklore

Sodium minerals have additionally tracked down their direction into fables and folklore. From antiquated accounts of the goddess of salt to stories of salt mines as supernatural domains, sodium minerals have been woven into the stories of various societies, mirroring their persevering through presence in the human creative mind.

3. **Purpose of the book: exploring sodium minerals geology and economic impact**

1. **Presentation**

The motivation behind this book is a convincing odyssey into the perplexing domains of sodium minerals, disentangling their geographical secrets and researching their significant monetary effect. As we set out on this investigation, the all-encompassing objective is to overcome any barrier between logical comprehension and the more extensive ramifications of sodium minerals in forming our reality. By digging into their geography and financial importance, we expect to give a complete and sagacious excursion for perusers across different foundations.

2. **Characterizing the Reason**

**2.1 Topography: Stripping Back Earth's Layers**

The land investigation of sodium minerals frames the underpinning of this book. We look to uncover the secret complexities of their arrangement, analyzing the land processes that lead to the making of these significant stores. From the old cycles that molded sodium-rich developments to the contemporary geographical scenes facilitating these minerals, every section unfurls a layer of the World's set of experiences.

Understanding the geography of sodium minerals includes interpreting the complicated transaction of components, powers, and time that has added to their arrangement. The synthetic organization, precious stone designs, and land developments all wind around together a story that reaches out a long ways past a simple logical comprehension.

It disentangles the tale of Earth's advancement and the powerful cycles that have prompted the production of sodium-rich geographical fortunes.

**2.2 Financial Effect: Revealing Worldwide Importance**

The financial effect of sodium minerals is the second mainstay of our investigation. These minerals, frequently concealed underneath the World's surface, hold an exceptional effect on worldwide businesses, exchange, and economies. Researching their financial importance includes a fastidious assessment of market patterns, provincial elements, and the more extensive ramifications of their extraction and usage.

Sodium minerals have been essential to financial development since forever ago, from old shipping lanes revolved around salt to the advanced modern buildings based on antacid creation. The financial sections of this book unfurl the account of sodium minerals as monetary impetuses, investigating their effect on work, provincial turn of events, and their job in molding worldwide exchange elements.

3. **Exploring the Topographical Scene**

**3.1 Topography: The Inconspicuous Powers Molding Sodium Minerals**

The topographical investigation starts by drenching perusers in the powers that shape sodium minerals. We dive into the profundities of the Earth, investigating the synthetic responses, temperature varieties, and land time scales that add to the arrangement of sodium-rich stores. Contextual analyses of conspicuous sodium mineral areas all over the planet will expose the topographical variety and uniqueness of each store.

Understanding the geographical complexities permits us to see the value in the uniqueness of sodium minerals, from the huge salt pads to the unpredictable gem designs of sodium feldspar. The land story additionally reaches out to the ecological contemplations of sodium mineral extraction, as we examine the harmony between tackling geographical riches and protecting the sensitive biological systems from which sodium minerals are uncovered.

**3.2 Financial Effect: The Gradually expanding influence of Sodium Minerals**

Changing into the monetary domain, we investigate the far reaching influence of sodium minerals across enterprises and locales. Worldwide market patterns, request supply elements, and the international parts of sodium mining shape the financial scene. Contextual analyses of areas vigorously subject to sodium mineral extraction will give experiences into the mind boggling connection between land overflow and monetary soundness.

From the effect of sodium minerals on work rates in mining areas to their part in the advancement of foundation and the foundation of modern centers, the financial investigation clarifies the multi-layered elements of sodium minerals.

Difficulties and open doors in the sodium mineral market will be taken apart, giving an exhaustive perspective on the financial powers at play.

4. **Cooperative energies among Topography and Economy**

**4.1 Interconnected Stories**

The cooperative energy among topography and the economy arises as a focal subject. By understanding the topographical groundworks of sodium minerals, we gain experiences into the financial potential they harbor. At the same time, grasping the monetary effect permits us to see the value in the geographical intricacies that add to the overflow or shortage of sodium mineral stores.

The interconnected stories portray sodium minerals, where land processes and financial powers combine. It becomes apparent that the meaning of sodium minerals rises above logical interest; it is profoundly weaved with the back and forth movement of human social orders, economies, and the fragile harmony between bridling Earth's assets and protecting its environments.

# Chapter 1

## Sodium Minerals Chemistry And Characteristics

Sodium is a fundamental component that assumes a critical part in different natural and land processes. It is an exceptionally receptive soluble base metal with the image Na and nuclear number 11. While sodium is usually connected with table salt (sodium chloride), it likewise happens in different minerals, adding to the World's outside layer piece. This article investigates the science and attributes of sodium minerals, revealing insight into their development, properties, and importance.

### 1. Sodium in the World's Hull:

Sodium is the 6th most plentiful component in the World's hull, comprising roughly 2.8% by weight. It fundamentally happens as minerals, with the absolute most normal sodium-bearing minerals being feldspar, albite, and sodium carbonate. Understanding the conveyance and commonness of sodium minerals is fundamental for fathoming their job in topographical cycles.

1. **Feldspar:**
   Feldspar is a gathering of rock-shaping minerals that comprise most of the World's hull. Among them, plagioclase feldspars contain critical measures of sodium. Albite, a particular kind of plagioclase feldspar, is made out of sodium aluminum silicate. It is a significant part in numerous molten rocks, like stone and diorite, and is pivotal in the development of different mineral stores.
2. **Sodium Carbonate:**

Sodium carbonate, normally known as soft drink debris or trona, is another significant sodium mineral. It has various modern applications, including glass creation, cleansers, and substance producing. Trona stores, found in different areas around the world, are a huge wellspring of sodium carbonate. These stores are the aftereffect of the precipitation of sodium minerals from vanishing waterways throughout topographical time scales.

**II. Arrangement of Sodium Minerals:**

The development of sodium minerals is a complicated interaction impacted by topographical and ecological variables. Understanding the circumstances under which these minerals are framed gives bits of knowledge into Earth's set of experiences and the development of its outside layer.

1. **Molten Cycles:**

   Sodium-rich minerals are ordinarily connected with molten processes. During the cooling and cementing of magma, feldspar gems can shape, integrating sodium into their construction. The particular mineral collections rely upon elements like temperature, pressure, and the organization of the magma.

2. **Aqueous Cycles:**

   Aqueous cycles, including the flow of hot liquids through rocks, additionally add to the development of sodium minerals. These liquids can disintegrate sodium from existing minerals and accelerate new sodium-rich minerals under unambiguous circumstances. This interaction is frequently connected with metal stores and can prompt the arrangement of financially significant sodium-bearing minerals.

3. **Evaporite Stores:**

In parched locales, the vanishing of water from saline lakes or oceans can bring about the testimony of sodium minerals. This is clear in the arrangement of sodium carbonate stores like trona. The sluggish vanishing of water permits sodium particles to join with carbonate particles, shaping strong sodium carbonate minerals.

**III. Compound Properties of Sodium Minerals:**

Sodium minerals show particular compound properties that originate from the attributes of the sodium particle. Sodium is profoundly responsive, particularly with water, and its presence in minerals impacts their general substance conduct.

1. **Dissolvability:**

   Numerous sodium minerals are water-dissolvable to changing degrees. This dissolvability is clear in the quick disintegration of sodium chloride in water, a peculiarity generally saw in regular day to day existence. The solvency of sodium minerals has suggestions for their vehicle and rearrangement in geographical cycles.

2. **Reactivity:**

   Sodium is known for its reactivity, especially when presented to dampness. This reactivity is reflected in the way of behaving of sodium minerals, particularly with regards to molten and aqueous cycles. Sodium-rich minerals might go through adjustment responses, affecting the by and large mineralogical piece of rocks.

3. **Shading:**

The shading of sodium minerals can fluctuate, contingent upon the presence of contaminations. For instance, feldspar minerals might show a scope of varieties because of hints of different components in their gem structure. Understanding the hue of sodium minerals is fundamental for mineral ID and can give signs about the geographical circumstances under which they shaped.

**IV. Meaning of Sodium Minerals:**

Sodium minerals have critical ramifications for both geographical cycles and human exercises. Their overflow and circulation influence the World's hull arrangement, and their monetary significance stretches out to different modern applications.

1. **Land Importance:**

   The presence of sodium minerals in rocks gives significant data about the geographical history of an area. By concentrating on the mineral arrays and their dissemination, geologists can induce the circumstances under which these minerals shaped, adding to how we might interpret Earth's advancement.

2. **Modern Applications:**

   Sodium minerals are essential in various modern cycles. Sodium carbonate, for example, is a vital part in the development of glass, cleansers, and different synthetics. The extraction and handling of sodium minerals add to ventures that assume a significant part in present day culture.

3. **Natural Contemplations:**

The mining and extraction of sodium minerals, similar to any modern action, raise ecological contemplations. Understanding the natural effect of sodium mineral extraction is fundamental for feasible asset the board and limiting the environmental impression related with these cycles.

**V. Difficulties and Future Bearings:**

In spite of how we might interpret sodium minerals, there are difficulties and unanswered inquiries that warrant further investigation. Propels in scientific methods and interdisciplinary examination can add to a more extensive comprehension of sodium minerals and their job in Earth's cycles.

1. **Insightful Strategies:**

   Progressions in scientific strategies, like X-beam diffraction, electron microscopy, and spectroscopy, empower researchers to portray sodium minerals at the nuclear and sub-atomic levels. These methods give experiences into the gem construction, structure, and properties of sodium minerals, upgrading our capacity to concentrate on their way of behaving and development instruments.

2. **Interdisciplinary Exploration:**
   Interdisciplinary exploration including topography, science, and ecological science is fundamental for a comprehensive comprehension of sodium minerals. Cooperative endeavors can resolve complex inquiries connected with the development, dissemination, and effect of sodium minerals on Earth's frameworks.

3. **Feasible Asset The executives:**

As the interest for sodium minerals keeps on developing, maintainable asset the board rehearses become progressively significant. Analysts and industry partners should cooperate to foster harmless to the ecosystem extraction techniques and investigate elective wellsprings of sodium minerals to limit the natural effect.

**1.1Overview of sodium as an element**

Sodium, with the synthetic image Na and nuclear number 11, is an exceptionally responsive soluble base metal tracked down in Gathering 1, Period 3 of the occasional table. This article gives an extensive outline of sodium as a component, covering its qualities, properties, and applications, without explicitly diving into the point by point compound response with water.

1. **Actual Properties:**
1. **Nuclear Construction:**
   Sodium has a nuclear number of 11, showing that it has 11 protons in its core. Its electron setup is [Ne] $3s^1$, meaning it has one valence electron in its peripheral energy level. This solitary valence electron adds to sodium's trademark reactivity.
2. **Actual State:**
   At room temperature (around 20°C or 68°F), sodium is a delicate, shimmering white metal. It has a generally low softening place of 97.79°C (208.02°F) and a limit of 882.9°C (1621.22°F). Sodium is known for its pliability and flexibility, however it is seldom utilized in its unadulterated metallic structure because of its outrageous reactivity.
3. **Thickness and Conductivity:**

Sodium is light, with a thickness of roughly 0.97 grams per cubic centimeter. It is a phenomenal conveyor of power, both as a strong metal and in its liquid state. The high electrical conductivity is credited to the portability of the solitary valence electron, which can move unreservedly inside the metal grid.

**II. Compound Properties:**

1. **Reactivity:**
   Sodium is prestigious for its high reactivity, especially with water. This reactivity

originates from the straightforwardness with which sodium can lose its valence electron to accomplish a steady electron design. Be that as it may, in this outline, we won't dig into the particular compound response with water.

2. **Development of Mixtures:**

Sodium promptly frames compounds with different components, particularly incandescent lamp like chlorine and non-metals like oxygen. Normal sodium compounds incorporate sodium chloride ($NaCl$), sodium hydroxide ($NaOH$), sodium carbonate ($Na_2CO_3$), and sodium bicarbonate ($NaHCO_3$). These mixtures have different modern applications.

**III. Event and Overflow:**

Sodium is the 6th most bountiful component in the World's covering, comprising around 2.8% of its all out weight. It principally happens as minerals, with sodium-bearing feldspar being one of the most predominant. Also, sodium is plentiful in seawater, where it is around 30.6% of the particles present.

**IV. Organic Importance:**

Sodium assumes a vital part in natural frameworks, especially in the working of cells and the support of homeostasis. Sodium particles ($Na^+$) are fundamental for nerve motivation transmission, muscle constriction, and the guideline of osmotic equilibrium in cells. The sodium-potassium siphon effectively moves sodium particles out of cells and potassium particles into cells, keeping up with the electrochemical inclination vital for cell capabilities.

**V. Modern Applications:**
**Sodium and its mixtures have various modern applications:**

1. **Sodium Hydroxide ($NaOH$):**
   Ordinarily known as acidic pop, sodium hydroxide is utilized in the creation of different products, including paper, materials, cleansers, and cleansers. It is additionally utilized in the oil and food handling enterprises.

2. **Sodium Carbonate ($Na_2CO_3$):**
   Otherwise called soft drink debris, sodium carbonate is urgent in the assembling of glass, cleansers, and different synthetic compounds. It is an imperative natural substance in the glass business, adding to the lucidity and strength of glass items.

3. **Sodium Chloride ($NaCl$):**
   Table salt, sodium chloride, has far and wide applications in the food business as a flavoring and additive. In the synthetic business, it is utilized for the creation of chlorine and sodium hydroxide through the chlor-soluble base cycle.

4. **Sodium in Metallurgy:**

Sodium is utilized in specific metallurgical cycles, for example, the decrease of titanium tetrachloride to deliver titanium metal. Furthermore, sodium is utilized in the extraction of a few responsive metals from their minerals.

### VI. Difficulties and Contemplations:

Sodium's reactivity presents difficulties in taking care of and stockpiling. The profoundly exothermic response with water and the potential for start of hydrogen gas require cautious safety measures while working with sodium. Appropriate capacity conditions, for example, putting away sodium under dormant gases or in oil, are fundamental to forestall accidental responses.

With regards to human wellbeing, checking and directing sodium admission are basic contemplations. Over the top sodium utilization, frequently connected with handled food sources and a high-salt eating regimen, has been connected to different medical problems, including hypertension and cardiovascular infections.

### 1.2 Common sodium minerals and their chemical properties

Sodium is a fundamental component with assorted applications in different fields, from industry to science. In topographical settings, sodium is many times tracked down in the World's covering as minerals. These sodium minerals add to the general structure of rocks and assume a urgent part in land processes. This article investigates some normal sodium minerals, their events, and their synthetic properties, revealing insight into their importance in the World's outside.

### 1. Feldspar Gathering:

The feldspar bunch is perhaps of the most plentiful mineral gathering in the World's outside, and a few individuals from this gathering contain critical measures of sodium. Feldspar minerals are aluminosilicates, meaning they contain aluminum, silicon, and oxygen in their precious stone construction. The two essential sodium-rich feldspar minerals are albite and oligoclase.

### 1. Albite ($NaAlSi_3O_8$):

Albite is a plagioclase feldspar mineral that is wealthy in sodium. Its synthetic recipe, $NaAlSi_3O_8$, mirrors its structure, with sodium (Na), aluminum (Al), silicon (Si), and oxygen (O) as its significant constituents. Albite is normally tracked down in granitic shakes and is a fundamental part in the development of different mineral stores.

**Synthetic Properties:**

**Dissolvability**: Albite is by and large insoluble in water, as are most silicate minerals. Be that as it may, enduring cycles over the long run can prompt the steady breakdown of albite, delivering sodium particles into watery conditions.

**Reactivity:** While albite is generally steady under ordinary topographical circumstances, it can go through change responses within the sight of specific liquids. These responses might prompt the development of auxiliary minerals and the arrival of sodium particles.

2. **Oligoclase (Na(AlSi$_3$O$_8$) - (Na,Ca)(Al,Si)$_2$O$_8$):**

Oligoclase is another plagioclase feldspar that contains sodium. Its arrangement can change inside the plagioclase series, where calcium can fill in for sodium. The synthetic equation mirrors this fluctuation, with both sodium and calcium assuming a part in the mineral's design.

**Synthetic Properties:**

**Strong Arrangement:** Oligoclase frames a strong arrangement series with albite and anorthite, where the extents of sodium and calcium can differ. This strong arrangement conduct is normal in plagioclase feldspars and results in a scope of creations inside the mineral gathering.

**Arrangement:** Oligoclase is usually found in volcanic shakes like rocks and diorites. Its arrangement is related with the crystallization of magmas, where the presence of sodium impacts the mineral collection.

**II. Sodium Carbonate Minerals:**

Sodium carbonate minerals, frequently alluded to as natron, are a fundamental gathering of sodium-bearing minerals. They assume a huge part in geographical settings, especially in parched districts where evaporative cycles concentrate broke down particles. Some striking sodium carbonate minerals incorporate trona, natron, and nahcolite.

1. **Trona (Na$_3$(CO$_3$)(HCO$_3$)·2H$_2$O):**

   Trona is a sodium carbonate mineral that ordinarily happens as an evaporite store. Its compound recipe, Na$_3$(CO$_3$)(HCO$_3$)·2H$_2$O, demonstrates the presence of sodium, carbonate, bicarbonate, and water particles in its gem structure.

   **Compound Properties:**

   **Dissolvability:** Trona is modestly solvent in water. Its dissolvability is a significant figure the development of trona stores, which ordinarily happen in basic lake conditions where vanishing prompts the precipitation of sodium carbonate minerals.

   **Event:** Trona stores are found in different areas around the world, with critical events in the US, China, and Africa. These stores are financially significant as a wellspring of sodium carbonate for modern applications.

2. **Natron (Na$_2$CO$_3$·10H$_2$O):**

Natron is a normally happening hydrated sodium carbonate mineral. Its substance equation, Na$_2$CO$_3$·10H$_2$O, shows that it contains sodium, carbonate, and ten water particles in its precious stone construction.

**Compound Properties:**

**Hydration:** The presence of ten water particles in its design makes natron a hydrated mineral. This hydration is significant in its arrangement, frequently happening in basic lake conditions where high dissipation rates lead to the crystallization of natron.

**Authentic Importance:** Natron has verifiable importance as it was utilized in old times in the protection of mummies in Egypt. Its capacity to retain dampness made it a significant substance in the embalmment cycle.

### III. Sodium Chloride (Halite):

Sodium chloride, usually known as halite or table salt, is one of the most notable sodium minerals. While it is essentially perceived for its culinary purposes, halite likewise has topographical importance as a typical evaporite mineral.

### 1. Halite (NaCl):

Halite is a basic ionic compound made out of sodium ($Na^+$) and chloride ($Cl^-$) particles. Its synthetic recipe, NaCl, mirrors a coordinated proportion of sodium to chloride particles.

**Synthetic Properties:**

**Solvency:** Halite is profoundly dissolvable in water, and its disintegration in regular habitats is a typical cycle. The solvency of halite adds to the arrangement of salt pads and the advancement of saline conditions.

**Evaporite Arrangement:** Halite is habitually found in evaporite stores, where the dissipation of water prompts the precipitation of disintegrated sodium and chloride particles. These stores can frame in dry locales or in encased bowls where water has no outlet.

### IV. Sodium Silicate Minerals:

Sodium is likewise a part in different silicate minerals, which comprise a critical piece of the World's hull. These minerals frequently contain sodium in complex silicate structures, where silicon and oxygen structure the foundation of the mineral.

### 1. Jadeite ($NaAlSi_2O_6$):

Jadeite is a sodium-rich pyroxene mineral that is a part of the gemstone jade. Its synthetic equation, $NaAlSi_2O_6$, demonstrates the presence of sodium, aluminum, silicon, and oxygen in its gem structure.

**Synthetic Properties:**

**Mineralogy:** Jadeite is a critical mineral in transformative rocks, especially in high-strain and high-temperature conditions. It is related with subduction zones where maritime outside is exposed to extreme land conditions.

**Gemstone:** Jadeite is esteemed as a gemstone, and its elaborate use goes back hundreds of years. It is valued for its lively green tone and is frequently cut into mind boggling gems and models.

## 2. Nepheline ($Na_3KAl_4Si_4O_{16}$):

Nepheline is a sodium and potassium aluminosilicate mineral with the compound equation $Na_3KAl_4Si_4O_{16}$. It has a place with the gathering of feldspathoid minerals and is generally tracked down in silica-poor molten rocks.

**Synthetic Properties:**

**Strong Arrangement:** Nepheline frequently shapes strong arrangement series with other feldspathoids like leucite and kalsilite. The fluctuation in mineral sythesis inside these strong arrangements is affected by variables like temperature and strain during development.

**Event:** Nepheline is a typical constituent in soluble volcanic rocks, including syenites and phonolites. Presence is characteristic of explicit magmatic conditions favor the crystallization of nepheline-bearing minerals.

**V. Sodium Sulfate Minerals:**

Sodium sulfate minerals are sulfate salts that contain sodium cations. These minerals are frequently connected with evaporite stores and can have different hydration states.

### 1. Glauberite ($Na_2Ca(SO_4)_2$):

Glauberite is a sodium-calcium sulfate mineral with the synthetic recipe $Na_2Ca(SO_4)_2$. It is a piece of the anhydrous sulfate mineral gathering, and its name is gotten from
Johann Rudolf Glauber, a German-Dutch scientist.

**Substance Properties:**

Anhydrous Structure: Glauberite is an anhydrous sulfate, meaning it doesn't contain water atoms in its precious stone design. It structures under unambiguous geographical circumstances where sulfate-rich saline solutions or vanishing marine waters lead to the precipitation of the mineral.

**Development:** Glauberite is regularly found in sedimentary conditions, including saline lakes and evaporite stores. Its event is frequently connected with dry or semi-dry areas.

### 2. Mirabilite ($Na_2SO_4 \cdot 10H_2O$):

Mirabilite, otherwise called Glauber's salt, is a hydrated sodium sulfate mineral with the substance equation $Na_2SO_4 \cdot 10H_2O$. It has a place with the class of sulfate minerals known as sulfates of the natron bunch.

**Substance Properties:**

**Hydration:** Mirabilite is profoundly hydrated, containing ten water particles in its precious stone design. This hydration is a consequence of its development in saline conditions with high dissipation rates.

**Dissolvability:** Mirabilite is solvent in water, and its dissolvability increments with temperature. Now and again, it can frame as efflorescent stores on surfaces in dry locales as broken up minerals in groundwater are brought to the surface and consequently vanish.

### 1.3 Unique characteristics and applications of sodium minerals

Sodium minerals, including a different cluster of mixtures, gloat exceptional qualities that add to their inescapable applications across different ventures. These minerals, particular from natural sodium, offer steadiness and flexibility in applications going from development to medical services. This investigation will dive into the particular properties of key sodium minerals, their event in nature, extraction techniques, and the wide cluster of utilizations they support.

### Sodium Chloride (NaCl): Normal Table Salt

Sodium chloride, regularly known as table salt, stands apart as one of the most recognizable sodium minerals. This glasslike compound happens normally in immense stores or because of the vanishing of saltwater. The one of a kind cubic gem construction of sodium chloride adds to its trademark taste and dissolvability in water. Its dissolvability in water makes it a fundamental part for preparing food, and its high liquefying and edges of boiling over (801 degrees Celsius or 1473 degrees Fahrenheit and 1413 degrees Celsius or 2575 degrees Fahrenheit, separately) render it reasonable for different modern cycles.

Past its culinary purposes, sodium chloride assumes a crucial part in food conservation, utilized in techniques, for example, pickling and relieving. Furthermore, its application as a de-icing specialist on streets in cool environments improves street wellbeing during winter. The multi-layered nature of sodium chloride makes it a foundation in day to day existence, from the kitchen to fundamental framework upkeep.

### Sodium Carbonate (Na2CO3): Soft drink Debris

Sodium carbonate, generally alluded to as soft drink debris or washing pop, remains as one more huge sodium mineral with unmistakable properties and a large number of uses. Its alkalinity, which permits it to respond with acids and control pH levels, is a key component. Moreover, its hygroscopic nature, demonstrating a preparation to retain dampness from the air, adds to its flexible arrangement of properties.

One of the essential uses of sodium carbonate is in the assembling of glass. Its basic nature helps with bringing down the liquefying point of silica, a urgent move toward the glass creation process. Past glassmaking, sodium carbonate is a central part in the development of different synthetic substances, including sodium bicarbonate and sodium hydroxide. Its job reaches out to water treatment, where its alkalinity demonstrates significant in processes that require pH level changes.

### Sodium Bicarbonate (NaHCO3): Baking Pop

Sodium bicarbonate, regularly known as baking pop, is a sodium mineral particular for its soluble properties and various applications. Utilized widely in the culinary world, baking soft drink goes about as a raising specialist in baking, making batter ascend by creating carbon dioxide gas. Its flexibility reaches out past the kitchen, tracking down applications in cleaning and individual consideration items because of its gentle grating properties and capacity to kill scents.

In medical care, sodium bicarbonate fills in as a stomach settling agent, killing abundance stomach corrosive and giving help from acid reflux. Moreover, its part in clinical medicines, for example, intravenous organization to treat specific corrosive base lopsided characteristics, features the assorted utilizations of this sodium mineral.

### Sodium Nitrate (NaNO3): Chile Saltpeter

Sodium nitrate, otherwise called Chile saltpeter, separates itself among sodium minerals for its job in both farming and the creation of manures. It happens normally in parched areas, outstandingly in Chile, and its extraction includes the disintegration of nitrate-rich stores, trailed by precipitation.

A fundamental part of manures, sodium nitrate gives establishes a wellspring of nitrogen, an essential supplement for their development. Its application in horticulture adds to expanded crop yields, making it a fundamental asset in supporting worldwide food creation. Furthermore, sodium nitrate has verifiable importance, as it was a critical fixing in the assembling of black powder.

### Sodium Sulfate (Na2SO4): Glauber's Salt

Sodium sulfate, frequently alluded to as Glauber's salt, is a sodium mineral with exceptional properties and applications in different enterprises. It is tracked down in nature as the decahydrate mineral mirabilite, which structures in saline lakes and salt springs. Extraction strategies regularly include dissolving mirabilite and afterward accelerating sodium sulfate.

One of the noticeable utilizations of sodium sulfate is in the assembling of cleansers. It fills in as a handling help, assisting with separating oil and oil during the washing system. In the material business, sodium sulfate is used in coloring processes, going about as an evening out specialist to guarantee even variety circulation. Its job reaches out to the development of paper, glass, and materials, displaying its flexibility in modern applications.

# Chapter 2

## Geological Formation Of Sodium Deposits

Sodium, an essential salt metal, holds huge significance in different modern applications. Its land development, especially with regards to sodium stores, includes complex cycles impacted by assorted factors. Understanding these geographical complexities is vital for asset investigation, extraction, and practical usage. In this investigation, we will dig into the topographical parts of sodium store arrangement, exploring the circumstances, components, and key variables molding the event and circulation of sodium stores.

### Geographical Foundation

Sodium, comprising around 2.4% of the World's hull, appears in minerals like sodium chloride (halite), sodium carbonate (natron), and sodium sulfate (mirabilite). These minerals are crucial in the arrangement of sodium stores and are firmly connected with explicit land conditions.

1. **Halite Stores**

   Halite, or sodium chloride, is a conspicuous sodium mineral and an essential part of salt stores. Its land arrangement principally spins around the dissipation of saline water bodies, prompting salt precipitation. This cycle happens in different land settings, including old oceans, shut bowls, and salt pads.

   In areas with parched or semi-bone-dry environments, where dissipation outperforms precipitation, saline water bodies observer an expansion in salt fixation. As sodium chloride focus surpasses its dissolvability limit, halite gems begin to accelerate. Over land timescales, these gems gather, framing significant salt beds. Instances of such stores remember the Salinas Grandes for Argentina and the Salar de Uyuni in Bolivia.

2. **Sodium Carbonate Stores**

   Sodium carbonate, generally known as natron, is another vital sodium-bearing mineral. The geographical arrangement of sodium carbonate stores is frequently connected to soluble lake frameworks and volcanic conditions. Natron is regularly tracked down in sedimentary stores, frequently interbedded with minerals like trona and nahcolite.

Soluble lakes, portrayed by high pH values, give an ideal climate to the precipitation of sodium carbonate minerals. The interaction includes the convergence of sodium-rich groundwater into the lake, where vanishing and synthetic responses add to the arrangement of natron stores. Remarkably, the Natron Bowl in Tanzania, arranged in the East African Crack Valley, stands apart as a critical area famous for its sodium carbonate stores.

3. **Sodium Sulfate Stores**

Sodium sulfate, fundamentally as mirabilite, is tracked down in different geographical settings, including salt pads, playas, and saline lakes. The land arrangement of sodium sulfate stores frequently includes the communication of sulfate-rich water with evaporative cycles.

In parched areas, where occasional water bodies experience intermittent flooding and drying, sodium sulfate can accelerate during the drying stage. The development of these stores is firmly connected to the dissolvability of mirabilite at various temperatures, with the mineral taking shape out of arrangement as the water cools and vanishes. Instances of districts with huge sodium sulfate stores remember the Incomparable Salt Lake for the US and the Qaidam Bowl in China.

**Factors Affecting Sodium Store Development**

A few land, hydrological, and climatic variables assume critical parts in the development of sodium stores. A far reaching comprehension of these variables is fundamental for foreseeing the event and dissemination of sodium-bearing minerals. Key elements impacting sodium store arrangement include:

1. **Environment and Vanishing Rates**
   Bone-dry and semi-dry environments with high vanishing rates make great circumstances for sodium focus in water bodies, prompting the development of sodium stores. The harmony among vanishing and precipitation essentially impacts the salt fixation in these conditions.

2. **Geothermal Movement**
   Volcanic conditions and geothermal movement add to the convergence of sodium-rich liquids into water bodies. This deluge can impact the arrangement of sodium stores, particularly in districts with uplifted geothermal movement.

3. **Hydrological Conditions**
   Shut bowls, salt pads, and soluble lake frameworks give ideal circumstances to the collection and safeguarding of sodium minerals. The hydrological elements of these conditions assume a basic part in sodium store development.

4. **Structural Movement**

   Structural cycles, like the fracturing of mainland plates, can make bowls and melancholies that act as possible locales for the gathering of sodium stores. The topographical history of an area, molded by structural movement, impacts its reasonableness for sodium store arrangement.

5. **Source Rocks**

   The creation of source rocks is crucial in deciding the sodium content of groundwater, which, thusly, impacts the potential for sodium store arrangement. The mineralogical cosmetics of the source rocks assumes a significant part in molding the geochemical qualities of sodium-rich liquids.

6. **Diagenesis and Adjustment**

Post-depositional processes, for example, diagenesis and adjustment, can alter sodium stores over the long run. These cycles can modify the mineralogical structure and actual qualities of sodium stores, affecting their in general land cosmetics.

**2.1 Processes leading to the formation of sodium mineral deposits**

Sodium, a profoundly receptive salt metal, assumes a critical part in different modern applications. One captivating part of sodium is its event in mineral stores, which are the consequence of mind boggling geographical cycles unfurling more than huge number of years. This broad investigation dives into the geographical, synthetic, and ecological variables adding to the beginning of sodium mineral stores, giving an inside and out comprehension of the cycles in question.

1. **Presentation**

   Sodium mineral stores are fundamental geographical developments that contribute altogether to the worldwide sodium supply. Understanding the cycles prompting their development includes a multidisciplinary approach, enveloping geography, science, and natural science. This complete assessment expects to disentangle the complicated interaction of variables impacting the beginning of sodium mineral stores.

2. **Land Setting**

   **2.1. Evaporite Conditions**

   One of the essential land settings for sodium mineral testimony is evaporite conditions. These conditions are described by the dissipation of saline water bodies, bringing about the fixation and precipitation of broken up minerals. In districts with dry or semi-parched environments, where dissipation rates surpass precipitation, enormous inland bowls can create. These bowls become stores for collecting sodium-rich minerals throughout land time scales.

   As water vanishes from these bowls, sodium-rich minerals like halite ($NaCl$) and trona ($Na_3(CO_3)(HCO_3)\cdot2H_2O$) encourage, framing broad stores. The Incomparable Salt Lake in the US and the Salar de Uyuni in Bolivia act as current instances of conditions where sodium minerals are effectively shaping through evaporite processes.

**2.2. Aqueous Cycles**

Aqueous cycles additionally add to the development of sodium mineral stores. Aqueous liquids, advanced with sodium and different minerals, relocate through breaks in the World's covering. As these liquids ascend towards the surface, they might experience cooler circumstances, prompting the precipitation of sodium-bearing minerals.

The aqueous beginning of some sodium stores is obvious in the relationship of sodium minerals with other aqueous elements like underground aquifers and springs. The collaboration between hot liquids and host rocks can bring about the draining of sodium from minerals like feldspar, a typical part of many rocks. Resulting cooling and changes in pressure consider the precipitation of sodium minerals inside the topographical arrangements.

3. **Natural Cycles**

Intriguingly, natural cycles likewise assume a part in the development of sodium mineral stores. In specific saline conditions, microorganisms, for example, microbes and green growth add to the precipitation of sodium minerals. These microorganisms produce extracellular polymeric substances (EPS), which can go about as nucleation locales for mineral precipitation.

In hypersaline conditions, where sodium focuses are high, the movement of halophilic microorganisms can prompt the arrangement of microbial mats. These mats trap residue and minerals, including sodium-bearing mixtures, adding to the advancement of sodium mineral stores. This natural impact on mineral precipitation adds a layer of intricacy to how we might interpret sodium store development.

4. **Environment and Ecological Elements**

Environment and natural circumstances are basic determinants in the development of sodium mineral stores. Parched and semi-bone-dry environments make ideal circumstances for the improvement of evaporite stores. The shortage of precipitation and high vanishing rates bring about the fixation and precipitation of sodium minerals from saline water bodies.

Structural movement and geographical inspire additionally impact the openness and availability of sodium-rich rocks to surface cycles. The accessibility of sodium in the World's covering is a forerunner to its delivery into arrangement and ensuing precipitation as minerals. The interchange between environment, tectonics, and other ecological variables decides the area and degree of sodium mineral stores worldwide.

5. **Diagenesis and Internment Cycles**

Diagenesis, the physical and substance changes that happen in dregs as they are covered and go through lithification, is another cycle adding to the development of sodium mineral stores. Entombment of saline residue, particularly in dying down bowls, can prompt the fixation and precipitation of sodium minerals.

As dregs aggregate and go through diagenesis, pore liquids inside the silt might turn out to be progressively saline. The compaction of these silt and the removal of pore liquids can bring about the precipitation of sodium minerals. Diagenetic processes are vital in grasping the drawn out land history of sodium mineral stores and the elements that impact their turn of events.

6. **Financial Importance**

The financial meaning of sodium mineral stores couldn't possibly be more significant. Sodium carbonate (soft drink debris), sodium bicarbonate, and sodium chloride (table salt) are among the economically significant minerals got from sodium stores.

Soft drink debris, a vital result of sodium stores, is utilized in the development of glass, cleansers, and different synthetic compounds. Sodium bicarbonate tracks down applications in baking, drugs, and water treatment. Sodium chloride, a universal mineral, is fundamental for human utilization, compound cycles, and de-icing streets in colder environments. The monetary feasibility of sodium mineral stores has prompted broad mining tasks in different areas of the planet.

7. **Contextual investigations**

**7.1. Salar de Uyuni, Bolivia**

Salar de Uyuni, the world's biggest salt level, is a momentous illustration of a functioning sodium mineral store. Situated in the Andes of southwest Bolivia, this salt level is a leftover of an ancient lake that evaporated, abandoning tremendous scopes of sodium-rich salt covering. The statement cycle in Salar de Uyuni is progressing, as the advanced environment keeps on working with the focus and precipitation of sodium minerals.

**7.2. Catch 22 Bowl, USA**

The Catch 22 Bowl in the US is famous for its broad evaporite stores. Throughout geographical time, the bowl has seen the collection of thick groupings of salt, including halite and other sodium-bearing minerals. The land history of the Catch 22 Bowl gives experiences into the drawn out processes associated with sodium mineral affidavit in evaporite conditions.

8. **Future Possibilities and Difficulties**

Understanding the cycles prompting the development of sodium mineral stores is a unique field with progressing exploration and difficulties. High level logical methods, including geochemical examinations and isotopic investigations, add to a more nuanced comprehension of the circumstances under which sodium minerals hasten.

The reasonable double-dealing of sodium assets requires a harmony between monetary contemplations and natural effect. Mining tasks should be directed in light of natural stewardship, taking into account the likely biological impacts of huge scope extraction.

## 2.2 Identification and classification of sodium-rich geological formations

Sodium, a major salt metal, isn't just a significant component for different natural cycles yet in addition a fundamental ware in modern applications. Understanding and recognizing sodium-rich land arrangements are crucial for proficient asset usage and vital preparation. This investigation digs into the systems, topographical markers, and characterization plans utilized in the distinguishing proof and grouping of sodium-rich geographical developments.

1. **Presentation**

   Sodium-rich geographical developments envelop a different exhibit of stores, going from evaporite conditions and aqueous frameworks to sedimentary bowls and saline solution pools. Distinguishing proof and grouping of these arrangements are essential for geographical examination, asset evaluation, and supportable double-dealing. The interaction includes a mix of field perceptions, lab examinations, and trend setting innovations.

2. **Topographical Pointers and Field Perceptions**

   **2.1. Evaporite Conditions**

   Evaporite conditions are essential wellsprings of sodium mineral stores. Distinguishing these conditions includes perceiving trademark elements like salt pads, playas, and saline lakes. The presence of halite ($NaCl$), trona ($Na_3(CO_3)(HCO_3)\cdot 2H_2O$), and other sodium-rich minerals is characteristic of evaporite affidavit.

   Field perceptions in evaporite conditions frequently uncover unmistakable sedimentary designs, including evaporite pseudomorphs, salt polygons, and parching breaks. The tinge of the stones, frequently white or shades of red, can likewise imply the presence of sodium minerals.

   **2.2. Aqueous Frameworks**

   Distinguishing proof of sodium-rich aqueous frameworks requires a sharp comprehension of related topographical elements. Aqueous adjustment zones, portrayed by mineral gatherings like albite ($NaAlSi_3O_8$) and zeolites, can act as markers. Field perceptions may likewise incorporate the presence of underground aquifers, fountains, and adjusted have rocks.

   Mineralogical drafting is normal in aqueous frameworks, with sodium minerals frequently found nearer to the intensity source. The modification of essential minerals, like feldspar, to sodium-rich auxiliary minerals gives extra insights for ID.

   **2.3. Sedimentary Bowls**

   Sedimentary bowls facilitating sodium-rich stores show explicit attributes that guide in distinguishing proof. Layers containing evaporites, including halite and sodium carbonate, are normal in these bowls. Topographical designs, for example, anticlines and synclines may trap and think sodium minerals.

Examination of sedimentary successions through hands on work and center testing gives bits of knowledge into the depositional history of these bowls. The ID of trademark sedimentary designs, like cross-sheet material and rhythmites, helps with figuring out the elements of sodium-rich sedimentation.

## 2.4. Salt water Pools

In districts with subsurface brackish water pools, distinguishing proof frequently includes surface appearances like salt outsides and blossoming. Salt water leaks might prompt the arrangement of saline soils, and the presence of explicit halophytic vegetation can be demonstrative of sodium-rich brackish waters. Geophysical strategies, including ground-entering radar and resistivity reviews, help in finding subsurface salt water supplies.

## 3. Lab Investigations and Geochemical Procedures

### 3.1. Geochemical Examination

Geochemical examination is a vital part in the distinguishing proof and order of sodium-rich topographical developments. Geochemical overviews include the assortment of tests, including rocks, dregs, and liquids, for research facility examination. Inductively Coupled Plasma Mass Spectrometry (ICP-MS) and Nuclear Ingestion Spectroscopy (AAS) are generally utilized to decide the convergences of sodium and related components.

Geochemical marks, like high sodium-to-potassium proportions, are demonstrative of sodium-rich conditions. Furthermore, isotopic investigation, including stable isotopes of sodium, can give experiences into the beginning and cycles administering sodium mineral affidavit.

### 3.2. Petrographic Investigation

Tiny assessment of slim areas utilizing petrographic examination is instrumental in distinguishing sodium minerals. Optical properties, cleavage examples, and mineral affiliations help in distinctive sodium-bearing minerals from different constituents. Petrographic examination is especially helpful in portraying the surface and texture of sodium-rich rocks.

### 3.3. X-beam Diffraction (XRD) and Spectroscopy

X-beam Diffraction (XRD) is a strong strategy for recognizing minerals in view of their gem structure. Sodium minerals show unmistakable diffraction designs that can be utilized for mineralogical recognizable proof. Infrared (IR) and Raman spectroscopy give corresponding data on the atomic sythesis of sodium-bearing minerals.

These methods add to the definite mineralogical portrayal of sodium-rich land developments, permitting scientists to observe explicit mineral stages present.

## 4. Grouping Plans for Sodium-Rich Land Developments

### 4.1. Hereditary Grouping

Hereditary grouping sorts sodium-rich topographical developments in view of their starting point and arrangement processes. The essential classifications incorporate evaporite stores, aqueous stores, sedimentary bowl stores, and saline solution pool stores. Every classification envelops particular topographical settings and cycles, working with a more extensive comprehension of sodium mineral beginning.

## 4.2. Mineralogical Characterization

Mineralogical characterization centers around the predominant sodium-bearing minerals present in geographical arrangements. Halite-overwhelmed stores, trona-ruled stores, and sodium carbonate-overwhelmed stores are instances of mineralogical orders.

This plan supports portraying the mineralogical variety of sodium-rich stores and understanding the topographical circumstances inclining toward explicit mineral collections.

## 4.3. Provincial Order

Provincial order thinks about the geographic conveyance of sodium-rich land arrangements. Various areas might show extraordinary geographical settings and structural impacts, prompting varieties in sodium mineralization. Local arrangement plans assist analysts and asset organizers with understanding the worldwide dispersion of sodium-rich stores and the geographical elements affecting their event.

5. **Cutting edge innovations in Recognizable proof**

## 5.1. Remote Detecting

Remote detecting innovations, including satellite symbolism and hyperspectral imaging, give significant devices to recognizing sodium-rich geographical developments from a good ways. Explicit ghastly marks related with sodium minerals can be recognized, supporting the depiction of sodium-rich regions. The mix of remote detecting with Geographic Data Framework (GIS) innovation improves spatial examination and planning.

## 5.2. Geophysical Overviews

Geophysical overviews, for example, magnetometry and resistivity studies, are instrumental in finding subsurface sodium-rich stores. Varieties in electrical conductivity and attractive powerlessness related with sodium minerals can be distinguished, giving data about the size and profundity of stores. Ground-entering radar is powerful in depicting the subsurface design of evaporite stores.

## 5.3. 3D Geographical Demonstrating

The appearance of 3D land demonstrating permits analysts to incorporate different datasets and make nitty gritty portrayals of sodium-rich geographical arrangements. Joining field perceptions, lab examinations, and geophysical information, 3D models give an exhaustive comprehension of the spatial conveyance and calculation of sodium stores.

## 6. Difficulties and Future Headings

In spite of progressions in ID and arrangement strategies, challenges persevere in precisely describing sodium-rich land developments. Subsurface investigation, especially in profound sedimentary bowls and aqueous frameworks, stays a specialized test. Furthermore, the intricate transaction of land processes requires a multidisciplinary approach for a far reaching understanding.

Future bearings in research include the improvement of novel logical methods and the combination of information from different sources. Propels in AI and computerized reasoning might add to robotized mineral distinguishing proof and upgrade the productivity of asset investigation.

# Chapter 3

## Mining Techniques And Extraction Processes

Mining is a deep rooted practice that has been necessary to human development, giving fundamental unrefined components to apparatuses, development, and different enterprises. As innovation has progressed, so too have mining procedures and extraction processes. This exhaustive investigation digs into the advancement of mining, the assorted techniques utilized for asset extraction, and the ecological and mechanical contemplations molding the business.

1. **Presentation**

   Mining is the most common way of extricating significant minerals or other geographical materials from the World's covering. Throughout the long term, the interest for minerals has driven the advancement of modern mining procedures. From early manual techniques to cutting edge innovations, the development of mining mirrors the steadily developing requirement for regular assets to help modern, infrastructural, and mechanical headways.

2. **Authentic Viewpoints on Mining Methods**

   **2.1 Manual Mining**

   Early mining exercises were described by difficult work and crude instruments. Diggers utilized basic handheld executes like picks and digging tools to extricate minerals from surface stores or shallow subsurface layers. This technique was work concentrated, restricting the scale and productivity of asset extraction. Exploration methods for sodium minerals

   **2.2 Shaft Mining**

   The progress to screw mining denoted a huge headway. Vertical shafts were dove into the Earth to get to more profound mineral stores. This strategy permitted excavators to arrive at significant assets concealed underneath the surface, growing the extent of mining tasks.

   **2.3 Open-Pit Mining**

The coming of open-pit mining addressed a progressive shift. Rather than burrowing into the Earth, huge surface unearthings were made, uncovering mineral stores. This strategy expanded effectiveness and empowered the extraction of tremendous amounts of metal. In any case, it additionally presented ecological difficulties, including environment disturbance and scene adjustment.

3. **Current Mining Methods**

**3.1 Strip Mining**

Strip mining is a variety of open-pit digging basically utilized for extricating coal and phosphate. It includes the evacuation of overburden (soil, rock, and vegetation) to uncover the mineral store. When uncovered, the mineral is extricated, and the overburden is refilled into the mined region. This method boosts asset recuperation yet raises worries about environment interruption.

**3.2 Underground Mining**

Underground mining has developed into a profoundly complex strategy for getting to firmly established mineral stores. There are different subtypes of underground mining:

**3.2.1 Room and Support point Mining**

This technique includes making chambers (rooms) isolated by mainstays of immaculate material. Diggers separate the mineral from the rooms, passing on the points of support to help the overlying stone. While effective, it can prompt subsidence, causing surface unsettling influences.

**3.2.2 Sublevel Stoping**

Sublevel stoping is regularly utilized for steeply plunging metal bodies. It includes making flat passages (sublevels) at various profundities. Mineral is then impacted and pulled from the sublevels. This technique is appropriate for medium to huge scope mining tasks.

**3.2.3 Block Buckling**

Block buckling is utilized for enormous, second rate metal bodies. It includes sabotaging the metal body, permitting it to fall under its weight. The messed up metal is then removed from underground passages. Block buckling is a practical strategy yet presents difficulties in overseeing subsidence and ground control.

**3.3 Placer Mining**

Placer mining includes the extraction of minerals, typically weighty metals like gold, from alluvial stores. This technique depends on the normal cycles of water and gravity to think significant minerals. Sluicing, digging, and panning are normal methods utilized in placer mining.

**3.4 Mountain ridge Expulsion Mining**

In mountain ridge expulsion mining, the culmination or highest point edge of a mountain is taken out to get to coal creases. This method has been especially dubious because of its extreme ecological effect, including deforestation, natural surroundings obliteration, and water contamination.

4. **Extraction Cycles**

Mining is only the underlying move toward getting significant minerals. Extraction processes are utilized to isolate and refine these minerals into usable structures. The decision of extraction technique relies upon the sort of mineral and its focus inside the metal.

**4.1 Actual Techniques**

**4.1.1 Gravity Division**

Gravity division takes advantage of the distinctions in thickness between minerals. In processes like jigging, centrifugation, and spiraling, heavier minerals settle quicker, considering division from lighter minerals. This technique is especially successful for gold and platinum.

**4.1.2 Attractive Partition**

Attractive partition utilizes the attractive properties of specific minerals to isolate them from non-attractive materials. This is regularly utilized in the extraction of iron mineral. Attractive separators draw in and eliminate ferrous minerals, abandoning non-attractive materials.

**4.2 Synthetic Techniques**

**4.2.1 Store Draining**

Store draining is a typical technique for separating metals from second rate minerals. Squashed mineral is stacked in stores, and a draining arrangement (generally a feeble corrosive) is applied. The arrangement permeates through the pile, dissolving the ideal metals. The pregnant arrangement is then gathered and handled to remove the metal.

**4.2.2 Refining**

Refining includes warming the metal to high temperatures within the sight of a decreasing specialist, like coke. This cycle is utilized for removing metals like copper and iron from their minerals. The metal is isolated from pollutants, creating a liquid metal that can be additionally refined.

**4.2.3 Dissolvable Extraction**

Dissolvable extraction is utilized for the division of metals from arrangements. Natural solvents are utilized to extricate metals from watery arrangements, framing a metal-rich natural stage. This strategy is essential in refining processes for copper, uranium, and uncommon earth components.

**4.3 Biotechnological Strategies**

Biotechnological strategies include the utilization of microorganisms to remove or recuperate metals from minerals. Bioleaching, for instance, uses microbes to oxidize and solubilize metals, working with their extraction. This harmless to the ecosystem approach is building up some decent momentum in the mining business.

5. **Natural Contemplations and Difficulties**

While mining and extraction processes are fundamental for satisfying the needs of a developing populace and propelling innovations, they are not without natural difficulties. Key worries include:

**5.1 Deforestation and Environment Disturbance**

Open-pit and peak expulsion mining frequently involve getting immense regions free from vegetation, prompting deforestation and natural surroundings annihilation. This change of biological systems can significantly affect neighborhood greenery.

**5.2 Water Contamination**

Mining exercises can bring about the arrival of hurtful substances into water bodies. Corrosive mine seepage, a typical issue in mining regions, happens when sulfide minerals are presented to air and water, creating acidic overflow that can pollute streams and lakes.

**5.3 Air Contamination**

The arrival of particulate matter and discharges from mining activities adds to air contamination. Residue and poisons can adversely affect air quality, presenting dangers to both the climate and human wellbeing.

**5.4 Subsidence and Land Corruption**

Certain mining techniques, especially those including underground mining, can prompt subsidence, causing the land surface to sink. This peculiarity can bring about land corruption and posture dangers to foundation.

**5.5 Waste Age**

Mining creates enormous amounts of waste, including overburden, tailings, and waste stone. Ill-advised administration of mining waste can bring about soil disintegration, draining of unsafe substances, and long haul ecological debasement.

6. **Mechanical Developments and Future Patterns**

The mining business is developing, driven by mechanical advancements pointed toward limiting natural effect and further developing proficiency. A few remarkable patterns and headways include:

**6.1 Computerization and Mechanical technology**

The coordination of computerization and mechanical technology improves wellbeing and productivity in mining activities. Independent vehicles, drones, and automated frameworks are being utilized for investigation, unearthing, and transportation in mines.

**6.2 High level Sensors and Investigation**

The utilization of cutting edge sensors and investigation takes into account constant checking of mining tasks. These advancements assist with streamlining processes, lessen energy utilization, and improve asset recuperation.

**6.3 Supportable Mining Practices**

Supportable mining rehearses center around limiting the natural impression of mining activities. This incorporates the utilization of sustainable power sources, water reusing, and taking on round economy standards to diminish squander.

### 6.4 Remote ocean Mining

With propels in innovation, remote ocean mining is turning into a subject of interest. This includes removing minerals from the seabed, and keeping in mind that it presents new difficulties, it might offer options in contrast to customary mining ashore.

### 3.1 Exploration methods for sodium minerals

The investigation for sodium minerals is a mind boggling and deliberate cycle that includes a blend of geographical, geochemical, and geophysical strategies. Sodium minerals, like halite, trona, and sodium carbonate, are essential assets with assorted modern applications.

This investigation expects to disentangle the complexities of recognizing and finding sodium mineral stores underneath the World's surface, utilizing state of the art procedures and advancements to explore the subsurface maze.

1. **Presentation**

   Sodium minerals assume an imperative part in different enterprises, from substance assembling to food creation. Distinguishing practical stores is fundamental for supportable asset usage. Investigation strategies for sodium minerals are intended to pinpoint subsurface stores, considering land settings, mineralogical attributes, and ecological contemplations.

2. **Work area Studies and Remote Detecting**

   **2.1 Writing Survey**

   The investigation cycle frequently starts with a far reaching writing survey to figure out the geographical setting of the objective region. Verifiable mining records, geographical guides, and scholastic distributions give important bits of knowledge into the territorial geography and potential sodium mineral events.

   **2.2 Remote Detecting Innovations**

   Remote detecting advancements, like satellite symbolism and ethereal overviews, assume a vital part in the underlying period of investigation. These strategies permit geologists to recognize surface highlights characteristic of sodium mineralization. For instance, the particular unearthly mark of salt pads or evaporite stores can be recognized utilizing hyperspectral sensors.

   High level satellite-based sensors, like those with warm infrared abilities, can distinguish surface temperature varieties related with explicit minerals. The incorporation of remote detecting information with Geographic Data Framework (GIS) innovation empowers the making of point by point maps that guide resulting ground-based investigation.

3. **Geographical Planning and Surface Examining**
   **3.1 Field Land Planning**
   Field land planning includes nearby reviews to investigate rock outcrops, sedimentary arrangements, and primary elements. Geographical guides are made to depict the dispersion of various stone sorts and potential sodium-bearing arrangements.
   **3.2 Surface Inspecting**
   Surface inspecting is a central investigation strategy that includes gathering rock, soil, or dregs tests from the objective region. Geochemical examinations of these examples give important data about the presence and grouping of sodium minerals. Inductively Coupled Plasma Mass Spectrometry (ICP-MS) and X-beam fluorescence (XRF) are ordinarily involved procedures for essential examination. Specific examining in view of known signs of sodium mineralization, for example, the presence of halite outsides or blooming, can upgrade the effectiveness of the investigation cycle. These surface examples assist geologists with affirming the presence of sodium minerals and figure out the geographical setting in more detail.
4. **Geophysical Reviews**
   **4.1 Gravity and Attractive Reviews**
   Gravity and attractive overviews are geophysical techniques that give bits of knowledge into subsurface topographical designs. Thickness varieties related with various minerals, including sodium minerals, can be distinguished through gravity studies. Attractive overviews assist with recognizing attractive minerals related with specific sorts of sodium stores.
   By breaking down peculiarities in gravity and attractive information, geophysicists can deduce the presence of subsurface designs or salt bodies. These overviews are especially helpful in recognizing covered evaporite stores.
   **4.2 Seismic Studies**
   Seismic reviews include sending acoustic waves into the ground and recording the time it takes for the waves to return. Different stone sorts and designs influence the speed and bearing of seismic waves. Seismic studies are significant for depicting subsurface designs, including salt vaults and sedimentary bowls where sodium minerals might aggregate.
   **4.3 Electromagnetic (EM) Studies**
   EM studies measure the electrical conductivity of the subsurface. Certain sodium minerals, as trona, show particular electrical properties. EM studies can assist with recognizing conductive layers demonstrative of sodium-bearing developments.
5. **Borehole and Penetrating Procedures**
   **5.1 Center Penetrating**
   Center penetrating includes separating barrel shaped examples (centers) from

the subsurface. This technique gives point by point data about the land layers and considers an immediate assessment of mineralogy. Center boring is especially helpful for grasping the upward conveyance of sodium minerals inside the World's outside layer.

**5.2 Mud Logging**

Mud logging is a strategy utilized during penetrating tasks to dissect the cuttings brought to the surface. The arrangement of these cuttings can give continuous data about the presence of sodium minerals. Mud logging is effective for rapidly evaluating mineralization while penetrating.

6. **Downhole Geophysics**

Downhole geophysics includes sending sensors and instruments into boreholes to straightforwardly gather subsurface information. This strategy gives high-goal data about the topographical layers and can be essential for describing the profundity and degree of sodium mineral stores.

7. **Geochemical Examination of Liquids**

Liquid testing includes gathering water or saline solution tests from wells or boreholes. Geochemical examination of these liquids recognizes the organization of disintegrated minerals, including sodium compounds. This technique is especially applicable for evaluating the financial feasibility of sodium mineral extraction.

8. **Trend setting innovations in Sodium Mineral Investigation**
   **8.1 AI and Information Joining**

The joining of AI calculations with investigation information works with prescient displaying. These models can distinguish examples and connections in huge datasets, assisting geologists with coming to additional educated conclusions about the probability regarding sodium mineral stores in unambiguous regions.

**8.2 High level Spectroscopy Procedures**

High level spectroscopy procedures, for example, laser-actuated breakdown spectroscopy (LIBS) and compact X-beam diffraction (pXRD), offer in-situ mineral ID capacities. These advancements can be conveyed straightforwardly in the field, giving quick and exact mineralogical data.

9. **Natural Contemplations in Sodium Mineral Investigation**

Sodium mineral investigation, similar to any mining action, should focus on ecological contemplations. Maintainable investigation rehearses incorporate limiting surface unsettling influence, sticking to recovery principles, and utilizing innovations that lessen the natural effect of investigation exercises.

10. **Difficulties and Future Bearings**

Investigation for sodium minerals faces difficulties like the profundity of stores, complex topographical settings, and the requirement for maintainable practices.

Future bearings in sodium mineral investigation might include the proceeded with mix of cutting edge innovations, an emphasis on natural obligation, and the investigation of unpredictable sources, including remote ocean sodium stores.

### 3.2 Mining technologies and extraction processes

Mining advances and extraction processes have developed fundamentally throughout the long term, molding the scene of asset extraction and use. From old manual procedures to state of the art advancements, the mining business has gone through extraordinary changes. This investigation digs into the different exhibit of mining advancements, extraction processes, and their effect on asset improvement.

1. **Presentation**

   Mining, a foundation of human development, includes extricating significant minerals and assets from the World's outside. The advancement of mining innovations and extraction processes mirrors the resourcefulness and versatility of human social orders in satisfying the needs of different businesses. This investigation intends to unwind the mind boggling woven artwork of mining innovations, featuring both authentic methodologies and contemporary progressions.

2. **Verifiable Mining Advancements**

   **2.1 Manual Mining**

   The earliest type of mining included physical work utilizing crude instruments like picks, digging tools, and sledges. Excavators would extricate minerals from surface stores or shallow subsurface layers. While work serious, manual digging established the groundwork for asset extraction and gave fundamental materials to apparatuses, development, and early innovations.

   **2.2 Shaft Mining**

   The progress to screw mining denoted a critical headway. Vertical shafts were dove into the Earth to get to more profound mineral stores. This strategy permitted excavators to arrive at significant assets concealed underneath the surface, growing the scale and proficiency of mining activities.

   **2.3 Open-Pit Mining**

   The coming of open-pit mining addressed a progressive change in mining innovation. Enormous surface unearthings were made, uncovering mineral stores. Open-pit mining expanded productivity, empowering the extraction of immense amounts of mineral. In any case, it presented natural difficulties, including territory disturbance and scene adjustment.

3. **Current Mining Advances**

   **3.1 Surface Mining**

   **3.1.1 Strip Mining**

   Strip mining is a variety of open-pit digging normally utilized for extricating coal and phosphate. It includes eliminating overburden (soil, rock, and vegetation) to uncover the mineral store. When uncovered, the metal is separated, and the overburden is inlayed into the mined region. This strategy augments asset recuperation yet raises worries about biological system disturbance.

### 3.1.2 Mountain ridge Expulsion Mining

Mountain ridge expulsion mining includes eliminating the culmination or highest point edge of a mountain to get to coal creases. This disputable technique has serious ecological effects, including deforestation, living space annihilation, and water contamination.

### 3.2 Underground Mining

### 3.2.1 Room and Point of support Mining

In room and point of support mining, chambers (rooms) are made and isolated by mainstays of immaculate material. Diggers remove the mineral from the rooms, passing on the points of support to help the overlying stone. While effective, it can prompt subsidence, causing surface aggravations.

### 3.2.2 Sublevel Stoping

Sublevel stoping is utilized for steeply plunging metal bodies. It includes making level passages (sublevels) at various profundities. Metal is then impacted and pulled from the sublevels. This technique is appropriate for medium to huge scope mining activities.

### 3.2.3 Block Buckling

Block buckling is utilized for huge, second rate mineral bodies. It includes subverting the metal body, permitting it to implode under its weight. The wrecked mineral is then separated from underground passages. Block buckling is savvy yet presents difficulties in overseeing subsidence and ground control.

### 3.3 Placer Mining

Placer mining includes separating minerals, typically weighty metals like gold, from alluvial stores. This technique depends on the normal cycles of water and gravity to focus important minerals. Sluicing, digging, and panning are normal procedures utilized in placer mining.

## 4. Mining Extraction Cycles

### 4.1 Actual Strategies

### 4.1.1 Gravity Partition

Gravity partition takes advantage of contrasts in thickness between minerals. In processes like jigging, centrifugation, and spiraling, heavier minerals settle quicker, considering partition from lighter minerals. This technique is compelling for gold and platinum.

### 4.1.2 Attractive Detachment

Attractive detachment utilizes the attractive properties of specific minerals to isolate them from non-attractive materials. This is generally utilized in the extraction of iron metal. Attractive separators draw in and eliminate ferrous minerals, abandoning non-attractive materials.

### 4.2 Substance Strategies

### 4.2.1 Pile Filtering

Pile filtering is utilized for removing metals from poor quality minerals. Squashed metal is stacked in stores, and a draining arrangement (typically a powerless corrosive) is applied. The arrangement permeates through the load, dissolving the ideal metals. The pregnant arrangement is then gathered and handled to separate the metal.

### 4.2.2 Refining

Refining includes warming the mineral to high temperatures within the sight of a decreasing specialist, like coke. This cycle is utilized for separating metals like copper and iron from their minerals. The metal is isolated from contaminations, delivering a liquid metal that can be additionally refined.

### 4.2.3 Dissolvable Extraction

Dissolvable extraction is utilized for the partition of metals from arrangements. Natural solvents are utilized to remove metals from watery arrangements, framing a metal-rich natural stage. This technique is essential in refining processes for copper, uranium, and uncommon earth components.

### 4.3 Biotechnological Strategies

Biotechnological strategies include the utilization of microorganisms to remove or recuperate metals from minerals. Bioleaching, for instance, uses microbes to oxidize and solubilize metals, working with their extraction. This harmless to the ecosystem approach is building up some decent forward momentum in the mining business.

5. ### Trend setting innovations in Mining

### 5.1 Computerization and Mechanical technology

The coordination of robotization and advanced mechanics upgrades security and effectiveness in mining tasks. Independent vehicles, drones, and automated frameworks are being utilized for investigation, uncovering, and transportation in mines.

### 5.2 High level Sensors and Examination

The utilization of cutting edge sensors and examination considers ongoing checking of mining activities. These innovations assist with improving cycles, decrease energy utilization, and upgrade asset recuperation.

### 5.3 3D Geographical Demonstrating

3D geographical demonstrating permits scientists to coordinate different data-sets and make definite portrayals of mining regions. Consolidating field perceptions, lab examinations, and geophysical information, 3D models give a far reaching comprehension of the subsurface construction.

6. ### Difficulties and Future Bearings

### 6.1 Ecological Worries

Mining exercises frequently bring about ecological difficulties, including deforestation, living space disturbance, water contamination, and air contamination. Offsetting asset extraction with ecological preservation is a basic test confronting the business.

## 6.2 Maintainable Mining Practices

The reception of supportable digging rehearses is significant for limiting the natural impression of mining activities. This incorporates dependable waste administration, water protection, and the utilization of environmentally friendly power sources.

## 6.3 Remote ocean Mining

With mechanical progressions, remote ocean mining is turning into a subject of interest. This includes separating minerals from the seabed and presents both innovative difficulties and ecological worries.

## 6.4 Roundabout Economy in Mining

The idea of a roundabout economy, which underlines reusing and reusing materials, is building up some forward momentum in the mining business. Carrying out roundabout economy standards can lessen squander age and advance economical asset the executives.

### 3.3 Environmental considerations and sustainable mining practices

Mining, a foundation of modern turn of events, gives fundamental natural substances to different areas. Notwithstanding, the natural effect of mining exercises has raised worries about living space disturbance, water and air contamination, and biological system debasement. This investigation dives into the natural contemplations related with mining and the advancement of reasonable practices pointed toward relieving these effects.

1. **Presentation**

   Mining, while fundamental for satisfying worldwide needs for minerals and assets, has generally left a huge impression on the climate. The extraction of minerals, modification of scenes, and removal of waste have prompted biological lopsided characteristics and ecological corruption. Perceiving these difficulties, the mining business has been progressively centered around embracing reasonable practices to limit its effect on the climate and advance capable asset usage.

2. **Ecological Effect of Mining**

   **2.1 Natural surroundings Interruption and Deforestation**

   One of the essential ecological worries related with mining is territory disturbance and deforestation. Open-pit and strip mining tasks frequently require getting enormous regions free from vegetation, prompting the deficiency of biodiversity and discontinuity of environments. This disturbance can bring about the dislodging of untamed life and modify the regular equilibrium of nearby biological systems.

   **2.2 Water Contamination**

   Mining exercises can add to water contamination through different components.

Corrosive mine waste is a typical issue, happening when sulfide minerals in uncovered rock respond with air and water, creating acidic spillover.

This spillover can debase close by water bodies, influencing amphibian life and water quality. Moreover, the release of mining effluents containing weighty metals represents a danger to sea-going biological systems.

### 2.3 Air Contamination

The arrival of residue and particulate matter during mining tasks adds to air contamination. The impacting, uncovering, and transportation of materials create airborne poisons, influencing air quality in the encompassing regions. Respiratory issues and other wellbeing concerns can emerge for both natural life and close by networks.

### 2.4 Soil Disintegration and Sedimentation

Surface mining exercises, particularly in sloping or precipitous locales, can prompt soil disintegration and sedimentation. The expulsion of vegetation and the unsettling influence of soil layers make scenes vulnerable to disintegration. Dregs overflow into water bodies can debase water quality, hurt sea-going environments, and upset silt balance.

### 2.5 Waste Age and Tailings The executives

Mining tasks create significant measures of waste, including overburden, squander rock, and tailings. Ill-advised administration of mining waste can bring about soil disintegration, draining of destructive substances into the climate, and long haul natural debasement. Tailings, specifically, represent a critical test because of their capability to contain harmful components and synthetic compounds.

### 3. Reasonable Mining Practices

### 3.1 Natural Effect Evaluation (EIA) and Checking

Maintainable mining rehearses start with a far reaching Natural Effect Evaluation (EIA) before the commencement of any mining project. EIAs assess the expected ecological outcomes of mining exercises and propose moderation measures. Continuous natural checking during and in the wake of mining activities guarantees consistence with ecological guidelines and considers versatile administration.

### 3.2 Recovery and Territory Reclamation

Recovery includes reestablishing mined regions to a condition reasonable for their expected future use. This might incorporate regrading the scene, establishing local vegetation, and once again introducing natural life. Recovery endeavors expect to limit the drawn out effect of mining on biological systems and biodiversity, advancing the rebuilding of normal territories.

### 3.3 Water The board and Treatment

Carrying out compelling water the board rehearses is critical for reasonable mining. This includes catching and treating overflow water to forestall tainting of neighboring water bodies. Building sedimentation lakes, utilizing water treatment offices, and carrying out best administration rehearses add to mindful water the board in mining activities.

### 3.4 Residue Control and Air Quality Administration

Dust control measures, for example, utilizing water splashes, windbreaks, and residue authorities, assist with limiting airborne particulate matter during mining activities. Powerful air quality observing and the executives designs further diminish the effect of mining exercises on neighborhood air quality and human wellbeing.

### 3.5 Waste Decrease and Reusing

Limiting waste age is a vital part of reasonable mining rehearses. This includes advancing mineral extraction cycles to lessen how much waste produced. Furthermore, investigating open doors for reusing and reusing materials inside the mining tasks can add to squander decrease.

### 3.6 Advancements in Tailings The executives

Creating imaginative ways to deal with tailings the executives is a point of convergence of practical mining. Innovations like sifted tailings, dry stacking, and glue refilling plan to lessen the ecological dangers related with customary tailings storage spaces. These advancements improve the regulation and the executives of tailings, moderating the potential for ecological mischief.

### 3.7 Local area Commitment and Social Obligation

Maintainable mining rehearses stretch out past natural contemplations to incorporate social obligation. Drawing in with nearby networks, tending to their interests, and adding to local area advancement projects cultivate positive connections between mining organizations and the networks they work in. This approach lines up with the standards of corporate social obligation (CSR) in the mining business.

### 4. Mechanical Developments in Reasonable Mining

### 4.1 Remote Detecting and Checking Advancements

Progressions in remote detecting advances, including satellite symbolism and automated ethereal vehicles (UAVs), empower constant checking of mining locales. These advances consider the location of changes in land use, vegetation wellbeing, and ecological circumstances. Remote detecting supports early ID of possible ecological issues, working with convenient intercessions.

### 4.2 Blockchain Innovation for Inventory network Straightforwardness

Blockchain innovation is being investigated to improve straightforwardness in the mining store network. By making a changeless and decentralized record, blockchain can follow the beginning of minerals, guaranteeing that they are obtained capably and morally. This innovation can possibly decrease the natural and social effect of mining by advancing capable obtaining.

### 4.3 High level Examination and AI

Progressed examination and AI calculations are progressively being applied in mining activities to upgrade cycles and asset use. These innovations break down huge datasets, giving bits of knowledge into energy effectiveness, squander decrease, and natural effect. Prescient demonstrating expects possible natural difficulties and guide decision-production for manageable results.

**4.4 Sustainable power Reconciliation**

Mining activities generally depend on huge energy utilization, frequently obtained from non-sustainable petroleum derivatives. Incorporating environmentally friendly power sources, for example, sun oriented and wind power, into mining tasks diminishes fossil fuel byproducts and decreases the ecological effect. Half breed power frameworks consolidating renewables with energy capacity innovations further add to supportable
energy rehearses.

5. **Certificate and Norms in Feasible Mining**

   **5.1 ISO 14001 Certificate**

   The Global Association for Normalization (ISO) gives the ISO 14001 certificate to ecological administration frameworks. Mining organizations can embrace this accreditation to show their obligation to manageable natural practices. ISO 14001 layouts a structure for executing compelling natural administration frameworks, advancing constant improvement, and guaranteeing consistence with ecological guidelines.

   **5.2 Towards Practical Mining (TSM) Drive**

   The Towards Reasonable Mining (TSM) drive, created by the Mining Relationship of Canada, gives a bunch of core values and execution pointers for capable mining rehearses. TSM covers regions, for example, tailings the board, energy use, biodiversity protection, and local area commitment. Mining organizations partaking in TSM focus on persistent improvement in these key regions.

6. **Difficulties and Future Headings**

   **6.1 Adjusting Monetary and Ecological Objectives**

   One of the continuous difficulties in practical mining is tracking down the right harmony between monetary targets and natural conservation. The requirement for mineral assets to help financial advancement should be weighed against the expected ecological effect, requiring cautious thought and inventive arrangements.

   **6.2 Worldwide Normalization and Requirement**

   While drives like ISO 14001 and TSM give systems to feasible practices, worldwide normalization and predictable requirement of ecological guidelines remain difficulties. Orchestrating principles and guaranteeing consistence across borders are critical stages in advancing capable mining rehearses on a worldwide scale.

   **6.3 Mechanical Openness for Limited scope Mining**

   Huge mining organizations frequently have the assets to put resources into trend setting innovations for economical practices. Notwithstanding, limited scope and distinctive mining tasks might confront provokes in taking on these advancements because of restricted monetary assets and mechanical openness. Finding savvy answers for more modest administrators is fundamental for inescapable reception of maintainable practices.

### 6.4 Advancement in Conclusion and Restoration

Compelling conclusion and restoration of mine destinations are necessary to manageable mining rehearses. Advancements in conclusion advances, like involving local plant species for recovery and carrying out long haul checking frameworks, are regions that require progressing innovative work.

# Chapter 4

## Applications Across Industries

Sodium minerals, including yet not restricted to halite, trona, and sodium carbonate, assume essential parts across different businesses because of their special properties and adaptable applications. From fundamental purposes in the food business to commitments in assembling, medical care, and natural areas, sodium minerals have become crucial unrefined components. This investigation digs into the assorted uses of sodium minerals across businesses, accentuating their financial importance and the effect on daily existence.

1. **Presentation**

   Sodium minerals, plentiful in nature, have been tackled by humankind for centuries. The one of a kind substance properties of sodium intensifies make them significant across a range of ventures, impacting item improvement, fabricating processes, and innovative progressions. This extensive investigation plans to disentangle the horde utilizations of sodium minerals, displaying their significance in molding different areas.

2. **Food Industry**

   **2.1 Table Salt (Halite)**

   One of the most natural uses of sodium minerals is in the development of table salt, basically got from halite. Sodium chloride ($NaCl$), the fundamental part of table salt, is an essential flavoring specialist used to improve the kind of different dishes. Past its job as a fixing, salt goes about as an additive, assisting with expanding the timeframe of realistic usability of food items.

   **2.2 Food Handling (Sodium Carbonate)**

   Sodium carbonate, ordinarily known as soft drink debris, finds applications in the food business, especially in food handling. It is utilized as a pH controller and a buffering specialist, helping control corrosiveness in different food items.

Sodium carbonate likewise adds to the raising system in baking, affecting the surface and nature of heated products.

3. **Synthetic Industry**

**3.1 Soft drink Debris (Sodium Carbonate)**

Soft drink debris, a sodium carbonate subsidiary, is an essential compound in the synthetic business. Its essential application lies in the development of glass. Sodium carbonate brings down the dissolving reason behind silica, working with the glass-shaping interaction. Glass fabricating, including the development of holders, windows, and fiber optics, intensely depends on soft drink debris.

**3.2 Sodium Bicarbonate**

Sodium bicarbonate, ordinarily known as baking pop, is a flexible compound with applications in the substance business. It is utilized in the creation of baking powder, which fills in as a raising specialist in baking. Past the culinary domain, sodium bicarbonate tracks down use in different compound cycles, including the treatment of wastewater and the drug business.

4. **Medical care and Drugs**

**4.1 Stomach settling agents and Sodium Bicarbonate**

In the medical care area, sodium bicarbonate fills in as a stomach settling agent, giving help from acid reflux and acid reflux. Its basic properties kill over-abundance stomach corrosive, offering a typical solution for gastrointestinal un-easiness. Furthermore, sodium bicarbonate is utilized intravenously in clinical settings to treat specific metabolic circumstances.

**4.2 Drug Excipients**

Sodium minerals, especially sodium chloride, are utilized as excipients in drug plans. They act as building specialists, stabilizers, and constitution changing specialists in different drugs. Sodium chloride infusions, for example, are critical in intravenous treatments, guaranteeing isotonicity with human blood.

5. **Water Treatment and Mellowing**

**5.1 Water Mellowing (Sodium Chloride)**

Sodium chloride assumes a vital part in water relaxing cycles. In areas where water is wealthy in calcium and magnesium particles, the utilization of sodium chloride in water conditioners replaces these particles with sodium particles, forestalling the development of scale in lines and machines. Water relaxing adds to further developed proficiency and life span of water-related apparatuses.

**5.2 Wastewater Treatment (Sodium Carbonate)**

Sodium carbonate finds applications in wastewater treatment as a pH-changing and coagulating specialist. It helps with the precipitation of pollutions, working with their expulsion from wastewater. The antacid idea of sodium carbonate advances the productive treatment of acidic effluents, adding to ecological manageability.

6. **Assembling and Industry**

   **6.1 Cleanser Creation (Trona)**

   Trona, a sodium carbonate mineral, is a vital unrefined substance in the creation of cleansers. Its soluble properties improve the cleaning effectiveness of cleansers, adding to the expulsion of oil and stains. Trona-determined sodium carbonate is an imperative part in planning clothing and dishwashing cleansers.

   **6.2 Metallurgical Cycles (Sodium Carbonate and Sodium Chloride)**

   Sodium carbonate and sodium chloride track down applications in metallurgical cycles. In the development of metals, for example, aluminum, sodium carbonate is utilized as a motion to bring down the dissolving point of alumina. Sodium chloride, then again, is utilized in different metallurgical applications, including the extraction of metals from minerals and as an intensity move medium in specific cycles.

7. **Development and De-icing**

   **7.1 Development Industry (Sodium Carbonate)**

   In the development business, sodium carbonate is used in different limits. It fills in as a pH controller in concrete assembling, adding to the setting and solidifying processes. Furthermore, sodium carbonate tracks down use in mortar creation, upgrading functionality and sturdiness.

   **7.2 De-icing (Sodium Chloride)**

   Sodium chloride, as street salt, is a generally utilized de-icing specialist during cold weather months. Spread on streets and walkways, it brings down the edge of freezing over of water, forestalling the arrangement of ice and upgrading security for vehicular and passerby traffic.

8. **Energy Capacity and Batteries**

   **8.1 Sodium-particle Batteries**

   As the interest for energy capacity arrangements develops, sodium minerals are acquiring consideration in the advancement of sodium-particle batteries. These batteries, comparable to lithium-particle batteries, use sodium particles for energy capacity. Sodium-particle batteries present a promising other option, offering possible benefits regarding cost and accessibility of unrefined components.

9. **Agribusiness and Soil Alteration**

   **9.1 Sodium Nitrate**

   Sodium nitrate, a sodium compound, fills in as a nitrogen manure in farming. It gives establishes a promptly accessible wellspring of nitrogen, advancing development and improving harvest yield. Sodium nitrate is especially valuable in districts with nitrogen-lacking soils.

10. **Natural Applications**

    **10.1 Residue Control (Sodium Chloride)**

    Sodium chloride is utilized for dust control on streets, building destinations, and unpaved surfaces. Its hygroscopic nature draws in dampness, forestalling the suspension of residue particles in the air. This application adds to further developed air quality and mitigates the natural effect of particulate matter.

## 10.2 Carbon Catch (Sodium Carbonate)

Sodium carbonate has been investigated as a possible specialist for carbon catch and capacity (CCS). In specific cycles, sodium carbonate responds with carbon dioxide to frame stable carbonates, working with the catch of $CO_2$ outflows from modern sources.

## 11. Difficulties and Future Viewpoints

### 11.1 Manageability and Mindful Obtaining

As the interest for sodium minerals increments, guaranteeing supportable and mindful obtaining rehearses becomes principal. Tending to ecological and social effects related with digging tasks is essential for keeping up with the drawn out suitability of sodium mineral extraction.

### 11.2 Mechanical Advancements

Proceeded with mechanical developments in mining, extraction, and handling of sodium minerals are fundamental for further developing proficiency and limiting natural effect. Investigation of novel extraction strategies, manageable handling procedures, and eco-accommodating practices will add to the development of the sodium mining industry.

### 11.3 Elective Materials and Reusing

Investigation into elective materials and reusing strategies can offer feasible answers for diminish reliance on conventional sodium minerals. Investigating substitutes with lower natural effect and creating proficient reusing cycles will assume a huge part in tending to asset supportability.

### 4.1 Industrial utilization of sodium minerals

Sodium minerals, including halite (rock salt), trona, and sodium carbonate, have arisen as fundamental assets in different modern applications, assuming basic parts in areas going from assembling and synthetics to medical services and development. This investigation digs into the different modern usage of sodium minerals, featuring their importance, applications, and the developing scene of innovation and advancement.

### 1. Presentation

Sodium minerals, bountiful in nature, have been outfit by ventures for a really long time. Their extraordinary synthetic properties, cost-viability, and adaptability make them key parts in various modern cycles. From the central utilization of halite in preparing to the high level uses of trona and sodium carbonate in assembling, these minerals add to the monetary development and mechanical progressions of social orders around the world.

### 2. Sodium Minerals in the Substance Business

#### 2.1. Soft drink Debris (Sodium Carbonate)

**2.1.1 Glass Assembling:**

One of the essential modern uses of sodium carbonate, generally known as soft drink debris, is in the glass fabricating industry. Sodium carbonate is a pivotal fluxing specialist that lessens the softening reason behind silica, working with the development of glass. Glass items, including jugs, windows, and fiberglass, vigorously depend on the consideration of pop debris in the assembling system.

**2.1.2 Substance Blend:**

Soft drink debris is a flexible compound utilized in different substance blends. It fills in as a forerunner in the creation of sodium bicarbonate, sodium silicates, and other sodium compounds. These subsidiaries track down applications in assorted businesses, including drugs, cleansers, and materials.

**2.2. Sodium Bicarbonate**

**2.2.1 Drugs:**

Sodium bicarbonate, regularly known as baking pop, assumes an essential part in the drug business. It is utilized as an acid neutralizer to kill overabundance stomach corrosive, giving help from heartburn and indigestion. Moreover, sodium bicarbonate is utilized in the detailing of specific drugs and intravenous treatments.

**2.2.2 Baking Industry:**

In the food business, sodium bicarbonate is widely utilized as a raising specialist in baking. It responds with acidic parts, delivering carbon dioxide and making mixture rise. This raising activity adds to the surface and volume of heated products, making sodium bicarbonate a staple in the baking business.

3. **Sodium Minerals in the Food Business**

**3.1. Halite (Table Salt)**

**3.1.1 Flavoring:**

Halite, or table salt, is a foundation of the food business, filling in as an essential flavoring and flavor enhancer. Its capacity to upgrade the flavor of different dishes makes it a universal fixing in kitchens around the world. Past its culinary purposes, salt goes about as an additive, broadening the time span of usability of food items.

**3.1.2 Food Handling:**

Salt assumes an essential part in food handling, adding to cycles like restoring, pickling, and maturation. It supports safeguarding meats, fish, and vegetables, forestalling decay and upgrading the security and kind of handled food sources.

4. **Sodium Minerals in Medical care**

**4.1. Sodium Bicarbonate**

**4.1.1 Stomach settling agent Properties:**

The stomach settling agent properties of sodium bicarbonate make it a typical solution for gastrointestinal inconvenience. Killing abundance stomach corrosive, giving alleviation from conditions, for example, acid reflux and heartburn is utilized. Moreover, sodium bicarbonate is managed intravenously in clinical settings to treat specific metabolic circumstances.

**4.1.2 Drugs:**

Sodium bicarbonate fills in as an excipient in drug details. It capabilities as a building specialist, stabilizer, and constitution changing specialist in different meds. Sodium bicarbonate infusions are urgent in intravenous treatments, guaranteeing isotonicity with human blood.

5. **Sodium Minerals in Water Treatment**

**5.1. Sodium Chloride**

**5.1.1 Water Relaxing:**

Sodium chloride, normally utilized as salt, is a central member in water mellowing processes. In regions where water is wealthy in calcium and magnesium particles, water conditioners use sodium chloride to supplant these particles, forestalling the development of scale in lines and apparatuses. This adds to further developed proficiency and life span of water-related apparatuses.

**5.1.2 Wastewater Treatment:**

Sodium chloride likewise tracks down applications in wastewater treatment. Its utilization assists in lessening the freezing with pointing of water in de-icing applications, forestalling the arrangement of ice on streets and walkways during cold weather months. This application upgrades wellbeing for vehicular and walker traffic.

**5.2. Sodium Carbonate**

**5.2.1 pH Change:**

Sodium carbonate fills in as a pH-changing specialist in water treatment processes. It keeps up with the ideal pH levels in both drinking water and wastewater treatment. The antacid idea of sodium carbonate helps with the coagulation and precipitation of pollutions, adding to the viable treatment of wastewater.

6. **Sodium Minerals in Assembling and Industry**

**6.1. Trona**

**6.1.1 Cleanser Creation:**

Trona, a sodium carbonate mineral, is a vital natural substance in the development of cleansers. Its basic properties upgrade the cleaning productivity of cleansers, making them more viable in eliminating oil and stains. Trona-determined sodium carbonate is a fundamental part in figuring out clothing and dishwashing cleansers.

**6.1.2 Glass Assembling:**

Like normal soft drink debris, trona is used in the glass producing industry. Its sodium carbonate content adds to the bringing down of the liquefying reason behind silica, working with the glass-shaping cycle. The utilization of trona gives an elective wellspring of sodium carbonate with expected monetary and ecological benefits.

## 6.2. Metallurgical Cycles

### 6.2.1 Aluminum Creation:

Sodium carbonate is utilized as a motion in metallurgical cycles, especially in the creation of aluminum. It helps bring down the liquefying point of alumina, the unrefined substance for aluminum creation. This works with the extraction of aluminum from alumina in a more energy-productive way.

### 6.2.2 Intensity Move Medium:

Sodium chloride is utilized as an intensity move medium in different metallurgical cycles. Its capacity to retain and deliver heat makes it a significant part in specific modern applications, for example, the cooling and warming cycles engaged with metal purifying.

7. **Sodium Minerals in Development**

## 7.1. Sodium Carbonate

### 7.1.1 Concrete Assembling:

Sodium carbonate is utilized in concrete assembling as a pH controller. It adds to the setting and solidifying cycles of concrete, affecting its last properties. The utilization of sodium carbonate in concrete creation upgrades the usefulness and solidness of cement.

## 7.2. Halite (Rock Salt)

### 7.2.1 De-icing:

Halite, as rock salt, is broadly utilized as a de-icing specialist during cold weather months. Spread on streets and walkways, rock salt brings down the edge of freezing over of water, forestalling the arrangement of ice. This application is basic for guaranteeing safe travel and person on foot development in cool environments.

8. **Sodium Minerals in Energy Stockpiling**

## 8.1. Sodium-particle Batteries

As the interest for energy capacity arrangements develops, sodium minerals are acquiring consideration in the improvement of sodium-particle batteries. These batteries use sodium particles for energy capacity, offering an expected option in contrast to lithium-particle batteries. Sodium-particle batteries present benefits concerning cost and the accessibility of unrefined components, adding to the feasible advancement of energy stockpiling innovations.

9. **Sodium Minerals in Farming**

## 9.1. Sodium Nitrate

### 9.1.1 Compost:

Sodium nitrate, a sodium compound, is utilized as a nitrogen compost in horticulture. It furnishes plants with a promptly accessible wellspring of nitrogen, advancing development and upgrading crop yield. Sodium nitrate is especially helpful in locales with nitrogen-lacking soils, adding to manageable farming practices.

## 10. Sodium Minerals in Natural Applications

### 10.1. Sodium Chloride

### 10.1.1 Residue Control:

Sodium chloride is utilized for dust control on streets, building destinations, and unpaved surfaces. Its hygroscopic nature draws in dampness, forestalling the suspension of residue particles in the air. This application adds to further developed air quality and mitigates the ecological effect of particulate matter.

### 10.1.2 Carbon Catch:

Sodium carbonate has been investigated as an expected specialist for carbon catch and capacity (CCS). In specific cycles, sodium carbonate responds with carbon dioxide to frame stable carbonates, working with the catch of CO2 outflows from modern sources. This application tends to natural worries connected with ozone harming substance outflows.

## 11. Difficulties and Future Viewpoints

### 11.1. Supportability and Dependable Mining

As the interest for sodium minerals keeps on rising, guaranteeing feasible and dependable mining rehearses becomes basic. Tending to ecological and social effects related with digging tasks is vital for keeping up with the drawn out suitability of sodium mineral extraction.

### 11.2. Mechanical Advancements

Proceeded with mechanical advancements in mining, extraction, and handling of sodium minerals are fundamental for further developing productivity and limiting natural effect. Investigation of novel extraction strategies, manageable handling procedures, and eco-accommodating practices will add to the advancement of the sodium mining industry.

### 11.3. Elective Materials and Reusing

Investigation into elective materials and reusing techniques can offer maintainable answers for lessen reliance on conventional sodium minerals. Investigating substitutes with lower natural effect and creating proficient reusing cycles will assume a critical part in tending to asset manageability.

### 4.2 Sodium minerals in chemical and manufacturing industries

Sodium minerals, enveloping mixtures like halite (rock salt), trona, and sodium carbonate, assume critical parts in the synthetic and assembling areas. With their remarkable substance properties, these minerals act as adaptable unrefined components, adding to processes going from glass assembling to drug creation. This investigation digs into the broad utilizations of sodium minerals in compound and assembling enterprises, revealing insight into their importance, processes included, and the developing scene of advancement.

1. **Presentation**

Sodium minerals, plentiful in nature, have been saddled by humankind for a really long time, developing from customary applications to becoming necessary parts in cutting edge modern cycles. The synthetic and assembling areas depend on the special properties of sodium minerals for their different applications. From working with glass arrangement to filling in as key parts in drugs, the usage of sodium minerals is at the very front of modern advancement and financial turn of events.

2. **Sodium Minerals in the Substance Business**

   **2.1. Soft drink Debris (Sodium Carbonate)**

   **2.1.1 Glass Assembling:**

   Soft drink debris, or sodium carbonate, remains as a foundation in the substance business, particularly in glass producing. The cycle includes bringing down the softening mark of silica, the principal part of glass, through the expansion of pop debris. This decrease in dissolving point works with the combination of silica and different parts, bringing about the development of glass items like jugs, windows, and fiberglass.

   **2.1.2 Synthetic Union:**

   Sodium carbonate fills in as a flexible compound in synthetic union. It goes about as a forerunner in the creation of different sodium compounds. Through synthetic responses, sodium carbonate can be changed into sodium bicarbonate, sodium silicates, and different subordinates. These mixtures track down applications in different businesses, including drugs, cleansers, and materials.

   **2.2. Sodium Bicarbonate**

   **2.2.1 Drugs:**

   Sodium bicarbonate, normally known as baking pop, assumes a urgent part in the drug business. Its stomach settling agent properties make it a typical solution for gastrointestinal uneasiness, killing overabundance stomach corrosive. Moreover, sodium bicarbonate is utilized intravenously in clinical settings to treat specific metabolic circumstances.

   **2.2.2 Baking Industry:**

   In the food business, sodium bicarbonate is a staple in the baking area. Its raising properties make it a fundamental fixing in baking powder. At the point when sodium bicarbonate responds with acidic parts in batter or hitter, carbon dioxide is delivered, making the combination rise. This raising activity adds to the surface and volume of different heated merchandise.

3. **Sodium Minerals in Assembling**

   **3.1. Trona**

   **3.1.1 Cleanser Creation:**

   Trona, a sodium carbonate mineral, is a vital participant in the assembling of cleansers. Its antacid properties improve the cleaning productivity of cleansers, making them more powerful in eliminating oil and stains. Trona-determined sodium carbonate is an essential part in forming clothing and dishwashing cleansers.

### 3.1.2 Glass Assembling:

Like normal soft drink debris, trona finds application in the glass fabricating industry. The sodium carbonate content in trona adds to the bringing down of the liquefying reason behind silica, working with the glass-framing process. The utilization of trona gives an elective wellspring of sodium carbonate with expected financial and natural benefits.

### 3.2. Metallurgical Cycles

### 3.2.1 Aluminum Creation:

Sodium carbonate fills in as a motion in metallurgical cycles, especially in the development of aluminum. It assumes a part in bringing down the softening mark of alumina, the essential unrefined substance for aluminum creation. This works with the extraction of aluminum from alumina in a more energy-proficient way.

### 3.2.2 Intensity Move Medium:

Sodium chloride, normally known as table salt, is utilized as an intensity move medium in different metallurgical cycles. Its capacity to retain and deliver heat makes it an important part in specific modern applications, for example, the cooling and warming cycles engaged with metal refining.

4. **Sodium Minerals in the Drug Business**

### 4.1. Sodium Bicarbonate

### 4.1.1 Acid neutralizer Properties:

The acid neutralizer properties of sodium bicarbonate make it a broadly involved fixing in the drug business. Killing overabundance stomach corrosive, giving alleviation from conditions, for example, acid reflux and heartburn is used. Sodium bicarbonate is likewise utilized in the detailing of specific drugs.

### 4.1.2 Drugs Excipients:

Sodium minerals, especially sodium chloride, act as excipients in drug definitions. Excipients are idle substances added to drugs to upgrade their security, bioavailability, and different properties. Sodium chloride infusions, for instance, are urgent in intravenous treatments, guaranteeing isotonicity with human blood.

5. **Sodium Minerals in Development**

### 5.1. Sodium Carbonate

### 5.1.1 Concrete Assembling:

Sodium carbonate finds application in the development business, especially in concrete assembling. It fills in as a pH controller, impacting the setting and solidifying cycles of concrete. The consideration of sodium carbonate in concrete creation improves the functionality and solidness of concrete, adding to the development of strong designs.

## 5.2. Halite (Rock Salt)

### 5.2.1 De-icing:

Halite, as rock salt, assumes a basic part in de-icing applications, particularly during cold weather months. Spread on streets and walkways, rock salt brings down the edge of freezing over of water, forestalling the arrangement of ice. This application is essential for guaranteeing safe travel and person on foot development in cool environments.

## 6. Sodium Minerals in Energy Stockpiling

### 6.1. Sodium-particle Batteries

The mission for proficient energy stockpiling arrangements has brought sodium minerals into the spotlight, with an emphasis on sodium-particle batteries. These batteries, similar to lithium-particle batteries yet involving sodium particles for energy capacity, present a promising other option. Sodium-particle batteries offer likely benefits concerning cost and the accessibility of unrefined components, adding to the supportable improvement of energy stockpiling advancements.

## 7. Difficulties and Future Points of view

### 7.1. Maintainability and Capable Mining

As the interest for sodium minerals raises, guaranteeing feasible and capable mining rehearses becomes principal. Tending to ecological and social effects related with digging activities is vital for keeping up with the drawn out feasibility of sodium mineral extraction. Carrying out eco-accommodating advances and sticking to moral mining rehearses are necessary parts of the business' supportable future.

### 7.2. Mechanical Developments

The eventual fate of sodium minerals in substance and assembling enterprises relies on constant mechanical developments. Investigation of novel extraction strategies, supportable handling procedures, and eco-accommodating practices are fundamental for further developing effectiveness and limiting natural effect. Cutting edge innovations, for example, robotization and accuracy mining, can improve the business' general manageability.

### 7.3. Elective Materials and Reusing

Investigation into elective materials and reusing strategies is basic for practical asset the executives. Investigating substitutes with lower ecological effect and creating productive reusing cycles will assume a huge part in tending to asset maintainability. Furthermore, advancement in squander decrease and reusing advancements can limit the natural impression of sodium mineral extraction and handling.

### 4.3 Role of sodium minerals in technological advancements

Sodium minerals, enveloping a scope of mixtures like halite (rock salt), trona, and sodium carbonate, have played a urgent and developing job in mechanical progressions. From their verifiable use in essential applications like flavoring and protection to their joining into complex cycles across different enterprises, sodium minerals have been basic to the improvement of innovation.

This far reaching investigation digs into the diverse job of sodium minerals in mechanical headways, crossing applications in gadgets, energy capacity, and arising advancements that influence the extraordinary properties of these minerals.

1. **Presentation**

   Sodium minerals, plentiful in nature, have been fundamental to human progress for centuries. While at first perceived for their essential purposes like flavoring food, these minerals have become impetuses for innovative advancement. The exceptional compound properties of sodium minerals, like their dissolvability, conductivity, and reactivity, make them significant parts in different mechanical applications. This investigation expects to unwind the broad job of sodium minerals in forming mechanical headways across different areas.

2. **Sodium Minerals in Hardware**

   **2.1. Sodium in Semiconductor Assembling**

   **2.1.1 Silicon Wafer Creation:**

   Sodium compounds, especially sodium carbonate, assume a part in the assembling of silicon wafers for semiconductor gadgets. Silicon wafers are the fundamental material for the development of incorporated circuits (ICs) and other electronic parts. Sodium carbonate is utilized in the cleaning and scratching processes during the creation of silicon wafers, adding to the accuracy and nature of semiconductor producing.

   **2.1.2 Glass Substrates in Presentations:**

   The presentation business depends on glass substrates for different electronic gadgets, including cell phones, tablets, and level board shows. Sodium minerals, particularly as soft drink debris (sodium carbonate), are critical in the creation of great glass substrates. The utilization of sodium minerals in glass fabricating guarantees the solidness, straightforwardness, and electrical properties expected for present day electronic showcases.

   **2.2. Sodium-particle Batteries**

   **2.2.1 Arising as an Energy Stockpiling Arrangement:**

   Sodium-particle batteries have arisen as a promising option in contrast to conventional lithium-particle batteries in energy capacity applications. Sodium minerals, especially sodium particles, are key parts in the electrochemical cycles of these batteries. The possible benefits of sodium-particle batteries incorporate expense adequacy and the overflow of sodium assets, adding to the advancement of maintainable and versatile energy stockpiling innovations.

   **2.2.2 Benefits and Difficulties:**

   The usage of sodium minerals in sodium-particle batteries offers a few benefits.

Sodium is more plentiful and more affordable than lithium, tending to worries about asset accessibility and cost in the creation of batteries. Be that as it may, difficulties, for example, energy thickness generally execution actually should be addressed for sodium-particle batteries to become broad in electronic gadgets and electric vehicles.

3. **Sodium Minerals in Energy Transformation**

   **3.1. Sodium-cooled Atomic Reactors**

   **3.1.1 Proficient Intensity Move:**

   Sodium-cooled atomic reactors address a mechanical headway in the field of thermal power. Fluid sodium fills in as a coolant in these reactors, proficiently moving intensity away from the atomic center. The utilization of sodium improves the general proficiency and security of atomic reactors, adding to the advancement of cutting edge atomic power age frameworks.

   **3.1.2 Difficulties and Exploration:**

   While sodium-cooled reactors offer benefits as far as intensity move, difficulties, for example, the potential for substance reactivity with water and the arrangement of strong sodium intensifies should be tended to. Continuous exploration expects to streamline the plan and wellbeing elements of sodium-cooled atomic reactors for a more reasonable and secure thermal power future.

4. **Sodium Minerals in Cutting edge Materials**

   **4.1. Sodium Silicates in Nanotechnology**

   **4.1.1 Nanoparticle Union:**

   Sodium silicates, got from sodium minerals, are necessary to nanotechnology applications. They are utilized in the union of silica nanoparticles, which track down applications in different fields, including medication, hardware, and materials science. Silica nanoparticles, balanced out by sodium silicates, show exceptional properties that are tackled for drug conveyance, imaging, and the improvement of cutting edge materials.

   **4.1.2 Catalysis and Nanocomposites:**

   Sodium minerals, especially as sodium silicates, act as impetuses in nanoparticle blend. The controlled development and control of nanoparticles empower the production of nanocomposites with custom-made properties. These nanocomposites have applications in the improvement of cutting edge materials, like lightweight and high-strength primary parts.

5. **Sodium Minerals in Natural Advancements**

   **5.1. Carbon Catch and Capacity (CCS)**

   **5.1.1 Using Sodium Carbonate:**

   Sodium carbonate, a sodium mineral subordinate, has been investigated for its expected job in carbon catch and capacity (CCS) advancements. In specific cycles, sodium carbonate responds with carbon dioxide ($CO_2$) to frame stable carbonates. This response is important for continuous examination expecting to foster proficient and practical strategies for catching and alleviating CO2 outflows from modern sources.

**5.1.2 Reasonable Arrangements:**

The usage of sodium minerals in CCS advancements adds to maintainable answers for tending to environmental change. By catching and putting away CO2 outflows, these advances plan to decrease the effect of ozone harming substances on the climate. Continuous headways in CCS examination might prompt the boundless reception of sodium-based arrangements in carbon catch advancements.

6. **Difficulties and Future Headings**

## 6.1. Innovative Incorporation and Increasing

**6.1.1 Joining in Cutting edge Gadgets:**

As sodium minerals keep on assuming a huge part in mechanical headways, coordinating them into cutting edge gadgets presents difficulties. The similarity of sodium-based materials with the requesting necessities of current hardware and energy stockpiling gadgets requires cautious thought. Innovative work endeavors are centered around enhancing the coordination of sodium minerals in cutting edge mechanical applications.

**6.1.2 Increasing Sodium-particle Battery Creation:**

The turn of events and commercialization of sodium-particle batteries face difficulties connected with increasing creation. Accomplishing economies of scale and guaranteeing the large scale manufacturing of superior execution sodium-particle batteries are basic elements for their inescapable reception. Research drives are in progress to resolve issues like anode materials, energy thickness, and generally speaking battery execution.

## 6.2. Advancements in Mining and Extraction

**6.2.1 Reasonable Mining Practices:**

As the interest for sodium minerals develops, it is basic to guarantee reasonable mining rehearses. Developments in mining advancements, extraction cycles, and waste administration are fundamental for limiting the ecological effect of sodium mineral extraction. Economical mining rehearses add to the drawn out accessibility of sodium assets and backing capable asset the executives.

**6.2.2 Reusing and Round Economy:**

Advancements in reusing innovations and the improvement of a roundabout economy for sodium minerals can additionally upgrade manageability. Reusing techniques for sodium-particle batteries, for instance, can assist with recuperating important materials and lessen reliance on essential sources. Headways in reusing add to asset effectiveness and limit the ecological impression of sodium mineral use.

## 6.3. Research in Arising Advances

**6.3.1 Quantum Registering and Sodium-based Materials:**

The field of quantum registering is acquiring consideration for its capability to alter data handling. Research is investigating the utilization of sodium-based materials in quantum registering applications. The one of a kind electronic properties of sodium particles might offer benefits in the improvement of quantum bits (qubits) and quantum data capacity.

**6.3.2 Sodium-based Sensors and Gadgets:**

Sodium-based materials are being researched for their likely in creating sensors and gadgets with improved execution. Research in materials science means to use the exceptional properties of sodium minerals to make creative electronic parts and gadgets. Sodium-based sensors might track down applications in regions like ecological checking and medical services.

# Chapter 5

## Economic Impact And Market Trends

1. **Presentation**

   Monetary effect and market patterns are dynamic powers that shape the contemporary worldwide business environment. A significant comprehension of their mind boggling interchange is basic for partners going from organizations and policymakers to financial backers. This broad investigation will investigate the complex components of monetary effect and market patterns, unwinding their interconnected nature and digging into their suggestions across different areas.

2. **Financial Effect**

1. **Macroeconomic Markers**

   The bedrock of financial examination lies in macroeconomic markers, boss among them being the Total national output (Gross domestic product). Gross domestic product, including utilization, venture, government spending, and net commodities, fills in as a compass for a country's monetary wellbeing. This part will investigate worldwide Gross domestic product patterns and the territorial subtleties that highlight monetary abberations. Expansion, a basic financial component, will be taken apart to figure out its causes, results, and the significant pretended by national bank strategies in overseeing it. Moreover, an investigation of joblessness, its different sorts, and the cultural repercussions of high joblessness rates will give bits of knowledge into the more extensive monetary scene. The assessment of loan costs will enlighten their importance, the impact of national bank approaches, and the flowing consequences for getting, speculation, and purchaser spending.

2. **Globalization and Exchange**

In an undeniably interconnected world, global exchange is a foundation of monetary essentialness. This part will examine the significance of worldwide exchange, the effect of economic deals and levies, and how these elements impact homegrown ventures and customers.

Furthermore, the advancement of worldwide stockpile chains, their weaknesses, and the basic for flexibility in store network the board will be investigated to understand the more extensive financial repercussions of globalization.

### III. Market Patterns

1. **Innovative Headways**

   The 21st century is seeing an exceptional rush of mechanical progressions that is reshaping enterprises across the range. Computerized change, a main impetus behind this change in outlook, will be analyzed for its unavoidable effect on business tasks and proficiency. Man-made brainpower (simulated intelligence) and mechanization, with applications spreading over different areas, will be investigated for their suggestions on work markets, moral contemplations, and strategy reactions.

2. **Ecological, Social, and Administration (ESG) Elements**

   Supportability has turned into a point of convergence in the business scene, driven by a developing consciousness of natural issues. This part will dig into corporate supportability rehearses and their effect on buyer inclinations and financial backer choices. Social obligation, enveloping corporate social obligation (CSR) and social effect effective financial planning, will be investigated close by the basic part of administration, underscoring the significance of vigorous corporate administration, administrative structures, and board variety.

3. **Segment Movements**

   Segment moves, a quiet yet strong power, are changing the financial scene. A maturing populace, with its significant ramifications for medical care, benefits, and work markets, will be inspected for the two difficulties and valuable open doors. At the same time, fast urbanization will be dissected for its foundation requests, metropolitan arranging goals, and the monetary open doors it presents.

### IV. Cross-Cutting Topics

1. **Coronavirus Pandemic**

   The Coronavirus pandemic, a worldwide disruptor of remarkable extents, has made a permanent imprint on monetary designs. This part will examine the present moment and long haul monetary effects, changes in buyer conduct, and the speed increase of advanced change. Moreover, the interconnectedness of general wellbeing and financial security will be investigated, revealing insight into the job of worldwide participation in building versatile wellbeing frameworks.

2. **International Elements**

International elements, portrayed by profession wars, taxes, political shakiness, and clashes, add an extra layer of intricacy to monetary examination. This segment will examine the acceleration and de-heightening of exchange pressures, their effects on worldwide stockpile chains, and the job of global associations in compromise. The impact of political vulnerabilities on venture and the monetary repercussions of international areas of interest will be analyzed.

## V. Future Standpoint

### 1. Arising Advances

What's in store holds commitment and difficulties in equivalent measure, with arising advancements ready to rethink businesses. Blockchain and digital currencies, stretching out past their monetary roots, will be investigated for their applications, administrative difficulties, and possible effects on monetary frameworks. At the same time, the coming of 5G innovation guarantees an extraordinary effect on correspondence organizations, with sweeping ramifications for different ventures. This part will analyze the difficulties and concerns connected with 5G organization.

### 2. Feasible Practices

The basic for feasible practices is picking up speed, with the change towards round plans of action and the rising significance of environmentally friendly power sources. The roundabout economy, stressing waste decrease and asset proficiency, will be investigated close by the progressions in sustainable power advances. The cooperative endeavors expected for an economical future will be featured.

### 5.1 Global demand and market trends for sodium minerals

### 1. Presentation

Sodium minerals assume a significant part in different businesses, adding to the worldwide interest for assorted applications. This examination investigates the market patterns and factors driving the worldwide interest for sodium minerals. From its applications in synthetic substances and assembling to its job in the food business, sodium minerals have become fundamental. Understanding the elements of worldwide interest and market patterns is fundamental for organizations, financial backers, and policymakers the same.

### 2. Outline of Sodium Minerals

Sodium minerals envelop a scope of mixtures, with normal models including sodium chloride (table salt), sodium carbonate (soft drink debris), sodium hydroxide (burning pop), and sodium bicarbonate. These minerals are necessary to various modern cycles and buyer items, making them essential parts of the worldwide economy.

### 3. Market Patterns

1. **Compound Industry**
   **Soft drink Debris (Sodium Carbonate)**
   Soft drink debris, a key sodium mineral, is broadly utilized in the synthetic business. The interest for soft drink debris is impacted by its job in the assembling of glass, cleansers, and different synthetic compounds. The worldwide pattern toward expanded development exercises, especially in arising economies, fills the interest for glass, consequently affecting the interest for soft drink debris. Also, the development of the cleanser business, driven by changing ways of life and cleanliness mindfulness, adds to the rising interest for sodium carbonate.

   **Burning Pop (Sodium Hydroxide)**
   Harsh soft drink is a flexible substance with applications in businesses like materials, mash and paper, and water treatment. The paper and mash industry, specifically, depends vigorously on burning soft drink for different cycles. As worldwide paper utilization keeps on rising, driven by bundling and cleanliness needs, the interest for harsh soft drink stays powerful.

2. **Fabricating and Metallurgical Cycles**
   **Sodium in Metallurgy**
   Sodium tracks down application in metallurgical cycles, especially in the extraction of specific metals. For example, sodium is utilized in the decrease of titanium, a basic component in aviation and other elite execution enterprises. As interest for lightweight and solid materials expands, the interest for titanium and, thusly, sodium in metallurgical cycles is supposed to develop.

3. **Food Industry**
   **Table Salt (Sodium Chloride)**
   Table salt is a pervasive sodium mineral utilized in families as well as in the food handling industry. The worldwide pattern toward handled and comfort food varieties straightforwardly affects the interest for sodium chloride.
   Nonetheless, expanding consciousness of medical problems connected with high salt utilization has prompted a counter-pattern of decreased sodium content in handled food sources. This double powerful requires cautious checking for organizations working in the sodium chloride market.

4. **Drugs**

**Sodium Bicarbonate**
Sodium bicarbonate, ordinarily known as baking pop, has applications in the drug business. It is utilized in the definition of specific prescriptions and is likewise utilized in clinical medicines. The drug business' development, driven by variables like a maturing populace and progressions in medical care, impacts the interest for sodium bicarbonate.

**IV. Factors Impacting Worldwide Interest**

1. **Populace Development and Urbanization**
   The worldwide populace is on a consistent ascent, joined by fast urbanization. As additional individuals move to metropolitan regions, there is an expanded interest for different items, remembering those depending for sodium minerals. Urbanization powers development exercises, foundation improvement, and the assembling area, all of which add to the interest for sodium minerals.

2. **Industrialization and Monetary Turn of events**
   Nations going through industrialization and monetary improvement witness a flood popular for synthetic compounds, fabricating information sources, and fundamental products. Sodium minerals, being key to these cycles, experience elevated interest in such conditions. The industrialization of arising economies essentially influences the worldwide market for sodium minerals.

3. **Innovative Headways**

   Headways in innovation can prompt the disclosure of new applications for sodium minerals or more proficient extraction and handling techniques. Moreover, mechanical advancements in end-client enterprises, for example, the improvement of eco-accommodating assembling processes, may impact the interest for sodium minerals.

4. **Natural and Administrative Contemplations**


Expanding consciousness of natural issues and a developing accentuation on maintainability can influence the interest for sodium minerals. For instance, the shift towards cleaner creation processes and the decrease of hurtful side-effects might change the prerequisites for specific sodium minerals.

**V. Difficulties and Open doors**


1. **Store network Elements**
   The extraction and handling of sodium minerals are much of the time gathered in unambiguous districts, prompting production network weaknesses. Political shakiness, administrative changes, or cataclysmic events in these districts can disturb the worldwide stock of sodium minerals, presenting difficulties for businesses dependent on these assets.

2. **Wellbeing and Ecological Worries**
   As wellbeing and ecological mindfulness ascend, there is a developing worry about the effect of over the top sodium utilization on wellbeing and the climate. This has prompted drives advancing diminished sodium content in food items and expanded examination of modern cycles including sodium minerals. Organizations working in this area need to explore these worries and possibly investigate supportable practices.

3. **Development and Exploration Potential open doors**

The sodium minerals industry presents amazing open doors for advancement and exploration, especially in the improvement of more practical extraction strategies, the disclosure of novel applications, and the making of sans sodium or low-sodium choices. Organizations putting resources into innovative work can situate themselves at the very front of these amazing open doors.

**5.2 Economic significance in regions with notable sodium mineral deposits**

1. **Presentation**

   Sodium minerals, including a scope of mixtures, for example, sodium chloride, sodium carbonate, sodium hydroxide, and sodium bicarbonate, are significant assets with different applications across ventures. Districts invested with eminent sodium mineral stores experience a significant financial importance that reaches out past simple topographical overflow. This far reaching examination will dive into the complex monetary significance of locales plentiful in sodium minerals, investigating how these stores become impetuses for modern development, business potential open doors, worldwide exchange elements, and foundation improvement.

2. **Modern Scene and Business Open doors**

1. **Work Creation and Financial Soundness**

   Districts with huge sodium mineral stores become central focuses for the foundation of extraction and handling enterprises. The mining and handling of sodium minerals, for example, soft drink debris and harsh pop, make a gradually expanding influence in the neighborhood economy by creating business valuable open doors. Talented and untalented work is participated in different aspects of the extraction cycle, from mining to refining and transportation. This occupation creation adds to monetary steadiness, diminishes joblessness rates, and supports the general prosperity of the nearby populace.

2. **Downstream Businesses and Modern Groups**

   The financial meaning of sodium mineral stores is amplified by their job as unrefined components in downstream ventures. The substance area, especially dependent on sodium minerals, lays out areas of strength for an in these districts. Soft drink debris, for instance, is a critical part in the glass producing industry. Districts with bountiful soft drink debris stores frequently witness the development of modern bunches, where glass fabricating offices, substance plants, and related businesses blend. This bunching impact improves collaborations, decreases creation costs, and invigorates the development of interconnected businesses.

**III. Worldwide Exchange Elements**

1. **Send out Potential and Unfamiliar Trade Profit**

   Locales with striking sodium mineral stores become vital players in the worldwide exchange of sodium-determined items. The commodity of sodium minerals and their subordinates contributes fundamentally to the locale's unfamiliar trade income. Nations enriched with these assets frequently end up in essential situations in global exchange discussions, utilizing their significance as providers of fundamental natural substances to different enterprises around the world. The financial effect stretches out to exchange associations and coalitions, encouraging discretionary and monetary relations.

2. **Exchange Lopsided characteristics and Key Situating**

The overflow of sodium mineral stores can impact a locale's exchange balance. By sending out sodium-inferred items, districts might gather exchange excesses, reinforcing their monetary situation on the worldwide stage. This essential situating works with financial development as well as permits areas to successfully explore worldwide exchange elements more. Notwithstanding, it additionally acquaints difficulties related with exchange awkward nature, requiring cautious financial administration and vital preparation.

**IV. Foundation Improvement**

1. **Transportation Organizations and Coordinated factors**
   The extraction and handling of sodium minerals require strong foundation advancement. Districts with prominent sodium stores observer the development of transportation organizations, including streets, rail routes, and ports, to work with the effective development of natural substances from mining locales to handling plants and completed items to homegrown and worldwide business sectors. This foundation upholds the sodium mineral industry as well as upgrades generally speaking local availability.

2. **Innovative work Centers**

Financial flourishing in areas with critical sodium mineral stores frequently draws in interests in innovative work (Research and development). Organizations working here apportion assets to advancement, looking for more proficient extraction techniques, investigating novel applications for sodium minerals, and tending to natural worries. This emphasis on Research and development changes these districts into centers of mechanical progressions in the sodium mineral industry, encouraging scholarly capital and driving advancement.

**V. Natural and Administrative Contemplations**

1. **Manageable Mining Practices**
   The extraction of sodium minerals, similar to any mining action, raises ecological worries. Areas with critical stores face the test of offsetting monetary double-dealing with manageable practices to guarantee the drawn out reasonability of

their assets. Maintainable mining works on, including natural effect evaluations, recovery endeavors, and the reception of cleaner creation advances, are basic to moderate the ecological impression of sodium mining.

2. **Local area Commitment and Corporate Social Obligation (CSR)**

The financial meaning of sodium mineral stores reaches out past industry lines to the neighborhood networks. Organizations associated with extraction and handling frequently take part in local area improvement drives and CSR exercises. These remember speculations for schooling, medical care, and foundation projects, adding to a superior personal satisfaction for occupants. By encouraging positive associations with neighborhood networks, organizations improve their social permit to work and add to the general prosperity of the district.

## VI. Difficulties and Moderation Techniques

1. **Asset Consumption and Economical Administration**
   One of the essential difficulties looked by districts with outstanding sodium mineral stores is the likely exhaustion of these assets over the long haul. Practical asset the board is basic to adjusting the financial advantages of extraction with long haul ecological contemplations. Executing estimates, for example, capable mining rehearses, occasional asset evaluations, and recovery endeavors can moderate the effect of asset consumption.

2. **Market Unpredictability and Enhancement**
   The financial meaning of sodium minerals opens districts to showcase unpredictability. Changes in worldwide interest, international occasions, and monetary slumps can affect the market elements of sodium-determined items. To relieve these dangers, locales might investigate broadening procedures. Distinguishing new applications for sodium minerals, creating enterprises less powerless to showcase variances, and adjusting to changing customer inclinations are essential parts of a versatile financial system.

3. **Natural Guidelines and Consistence**

The extraction and handling of sodium minerals frequently face severe natural guidelines. Districts with huge stores should explore administrative systems to guarantee consistence. Putting resources into advancements that diminish natural effect, carrying out prescribed procedures in squander the executives, and participating in straightforward detailing can assist these districts with meeting administrative necessities while keeping a supportable and dependable way to deal with sodium mineral extraction.

## VII. Future Standpoint and Development

1. **Innovative Progressions and Industry 4.0**
   The future monetary meaning of areas with eminent sodium mineral stores is entwined with mechanical headways. Industry 4.0 innovations, like mechanization, man-made consciousness, and information investigation, can possibly alter the sodium mineral industry. Computerization in mining and handling can upgrade effectiveness, decrease costs, and further develop wellbeing. Developments in extraction techniques and handling innovations will be urgent for keeping up with seriousness in a quickly advancing worldwide scene.
2. **Roundabout Economy and Supportable Practices**

The arising accentuation on the roundabout economy gives potential open doors for areas sodium mineral stores to embrace economical practices. Reusing and reusing sodium-determined items, investigating shut circle frameworks, and limiting waste can adjust the business to round economy standards. Organizations and districts embracing economical practices are probably going to acquire an upper hand in a market where ecological contemplations progressively impact customer and financial backer inclinations.

**5.3Challenges and opportunities in the sodium mineral market**

1. **Presentation**
   The sodium mineral market, enveloping mixtures, for example, sodium chloride, sodium carbonate, sodium hydroxide, and sodium bicarbonate, is a significant area with different applications across ventures. Nonetheless, similar to any market, it isn't resistant to difficulties and amazing open doors. This investigation will dive into the multi-layered scene of the sodium mineral market, investigating the deterrents it faces and the expected roads for development and advancement.
2. **Challenges in the Sodium Mineral Market**
1. **Asset Consumption and Supportability**
   **Over-Extraction Concerns:** The extraction of sodium minerals from regular stores raises worries about over-extraction and exhaustion of these assets. Districts intensely dependent on sodium minerals should explore a fragile harmony between fulfilling current need and guaranteeing the drawn out supportability of their stores.
   **Natural Effect:** Sodium mining activities can have ecological repercussions, including living space interruption, water contamination, and soil debasement. Supportable mining rehearses and capable natural administration are basic to address these difficulties and limit the biological impression of sodium mineral extraction.
2. **Market Instability and Financial Variances**
   **Worldwide Interest Changeability:** The sodium mineral market is impacted

by worldwide interest elements, which can be unpredictable. Financial slumps, international strains, and changes in purchaser inclinations can affect the interest for sodium-determined items. Districts vigorously reliant upon sodium mining might confront monetary vulnerabilities during times of market slumps.

**Reliance on Unambiguous Ventures:** The market's weakness to variances in unambiguous enterprises, for example, glass assembling or synthetic compounds, represents a test. Over-dependence on specific areas can open locales to financial dangers when these ventures face slumps or interruptions.

3. **Administrative Consistence and Natural Guidelines**

    **Severe Natural Guidelines:** Legislatures overall are progressively forcing tough ecological guidelines on mining and modern exercises. Consistence with these norms adds functional intricacies and expenses for sodium mineral extraction and handling.

    **Social Permit to Work:** Getting and keeping a social permit to work is pivotal. Local area concerns with respect to the ecological effect of mining tasks can prompt administrative difficulties and public resistance. Organizations in the sodium mineral market need to put resources into local area commitment and manageable practices to tie down their social permit to work.

4. **Wellbeing and Security Concerns**

**Laborer Wellbeing:** Mining and handling sodium minerals imply inborn dangers to specialist security. Guaranteeing strong wellbeing and security conventions is fundamental to forestall mishaps, safeguard laborers, and keep up with functional congruity.

**Shopper Wellbeing Mindfulness:** Expanding buyer consciousness of medical problems connected with over the top sodium utilization presents difficulties for specific sodium-determined items, like table salt. Producers need to adjust to changing customer inclinations, including the interest for low-sodium or sans sodium options.

**III. Open doors in the Sodium Mineral Market**

1. **Mechanical Headways and Development**

    **High level Extraction Procedures:** Putting resources into cutting edge mining innovations and extraction techniques can improve proficiency, diminish natural effect, and expand the existence of sodium mineral stores. Advancements, for example, mechanized mining processes and harmless to the ecosystem extraction strategies offer open doors for market players.

    **Item Expansion:** Investigating novel applications for sodium minerals past customary purposes presents huge open doors. Innovative work endeavors can prompt the formation of new items, extending the market and relieving gambles related with reliance on unambiguous enterprises.

2. **Worldwide Manageability Drives**

   **Roundabout Economy Works on:** Embracing round economy standards, for example, reusing and reusing sodium-determined items, lines up with worldwide maintainability drives. Organizations taking on shut circle frameworks could diminish at any point squander as well as add to a more economical sodium mineral market.

   **Green Advances:** The interest for harmless to the ecosystem innovations is rising. Creating green cycles for sodium extraction and assembling can situate organizations as pioneers in manageable works on, meeting the developing inclinations of earth cognizant buyers and financial backers.

3. **International Vital Situating**

   **Expansion of Exchange Associations:** Areas plentiful in sodium mineral stores can decisively broaden their exchange organizations. Laying areas of strength for out with a scope of countries can relieve the effect of international strains and give dependability to sodium mineral commodity markets.

   **Interest in Framework:** Upgrading foundation, including transportation organizations and ports, can work on the productivity of sodium mineral commodities. Locales putting resources into current foundation can situate themselves as appealing exchange accomplices and draw in unfamiliar speculations.

4. **Buyer Training and Wellbeing Patterns**

   **Low-Sodium Options:** Answering the rising consciousness of medical problems connected with sodium utilization, the market has the amazing chance to create and advance low-sodium or without sodium choices. Buyer training efforts can assist with moving inclinations toward better choices.

   **Advancements in Food Industry:** Coordinated efforts with the food business to foster imaginative sodium decrease advances can set out new market open doors. Organizations zeroing in on better choices can benefit from advancing shopper inclinations for nutritious and adjusted consumes less calories.

5. **Innovative work Speculations**


**Supportable Practices:** Consistent interests in innovative work can prompt the revelation of reasonable practices in sodium mineral extraction and handling. From harmless to the ecosystem mining procedures to energy-effective assembling processes, Research and development can drive positive changes in the business.

**Enhanced Item Portfolio:** Organizations putting resources into innovative work can broaden their item portfolios. This mitigates gambles related with market instability as well as positions these organizations as trend-setters equipped for adjusting to changing business sector requests.

# Chapter 6

## Environmental Sustainability

Ecological maintainability alludes to the capable use and protection of the World's regular assets, biological systems, and biodiversity to guarantee that current and people in the future can address their issues. This idea is established in the comprehension that human exercises significantly affect the climate and that it is vital for balance financial, social, and ecological contemplations for the prosperity of the planet.

### Presentation: Grasping Natural Maintainability

### Characterizing Natural Manageability

Natural manageability is a multi-layered idea that envelops different standards, including protection, recovery, and mindful asset the executives. At its center, it includes simply deciding and making moves that limit the adverse consequence of human exercises on the climate, advancing environmental equilibrium, and guaranteeing the drawn out strength of the planet.

### The Interconnectedness of Earth's Frameworks

One principal part of ecological supportability is perceiving the interconnectedness of Earth's frameworks. The climate, hydrosphere, lithosphere, and biosphere are complicatedly connected, and disturbances in a single framework can have expansive ramifications for other people. Understanding these interconnections is significant for concocting viable manageability techniques.

### Verifiable Point of view: Development of Natural Mindfulness

### Early Ecological Worries

While the idea of natural supportability has acquired conspicuousness in ongoing many years, authentic proof recommends that social orders since the beginning of time knew about the outcomes of ecological corruption. Early human advancements frequently wrestled with soil disintegration, deforestation, and water contamination, prompting the breakdown of certain social orders.

### Current Ecological Development

The twentieth century saw the rise of the cutting edge natural development. Occasions, for example, the distribution of Rachel Carson's "Quiet Spring" during the 1960s and the primary Earth Day in 1970 checked urgent minutes in raising public mindfulness about natural issues. These occasions laid the foundation for ensuing ecological regulation and global collaboration on maintainability.

### Difficulties to Natural Supportability

### Populace Development and Asset Consumption

One of the essential difficulties to natural manageability is the dramatic development of the worldwide populace. As additional individuals possess the planet, the interest for assets, like food, water, and energy, increments. This puts huge tension on environments and can prompt over-double-dealing and consumption of normal assets.

### Industrialization and Contamination

The quick industrialization of the previous century has achieved exceptional monetary development yet at the expense of ecological corruption. Modern exercises discharge contaminations very high, water, and soil, adding to environmental change, air contamination, and the downfall of biodiversity. Offsetting modern improvement with ecological insurance is an intricate test.

### Environmental Change and A worldwide temperature alteration

Environmental change, driven basically by human exercises like the consuming of petroleum derivatives, deforestation, and modern cycles, represents a serious danger to natural manageability. Climbing worldwide temperatures, outrageous climate occasions, and interruptions to biological systems are indications of environmental change. Relieving and adjusting to these progressions are basic parts of supportability endeavors.

### Loss of Biodiversity

The continuous loss of biodiversity is a squeezing worry for ecological supportability. Human exercises, including environment obliteration, contamination, and environmental change, add to the elimination of species at a disturbing rate. Biodiversity misfortune dissolves the planet's regular excellence as well as upsets biological systems and can have flowing consequences for human social orders.

### Standards of Ecological Supportability

1. **Protection of Assets**

   Protection includes utilizing assets carefully and keeping away from inefficient practices. This rule underscores the significance of decreasing utilization, reusing materials, and reusing to expand the life expectancy of assets and limit ecological effect.

2. **Sustainable power Sources**

   Progressing from petroleum products to sustainable power sources is a critical component of natural supportability. Sunlight based, wind, hydroelectric, and other environmentally friendly power advancements offer cleaner options in contrast to conventional energy sources, diminishing ozone harming substance outflows and moderating environmental change.

3. **Environment Rebuilding**

Endeavors to reestablish and restore biological systems that have been harmed or debased are fundamental to ecological supportability. This incorporates reforestation, wetland reclamation, and measures to safeguard and improve biodiversity.

4. **Supportable Agribusiness and Food Frameworks**

Horticulture is a huge supporter of ecological effect, from deforestation for farmland to the utilization of compound manures. Supportable horticulture centers around rehearses that advance soil wellbeing, water protection, and biodiversity while guaranteeing food security for a developing worldwide populace.

5. **Round Economy**

The idea of a round economy spins around limiting waste and boosting the reuse and reusing of materials. This approach expects to split away from the direct "take-make-arrange" model and make a framework where assets are constantly cycled to decrease natural effect.

6. **Dependable Utilization and Creation**

Empowering dependable utilization designs includes advancing items with insignificant natural effect, supporting moral and economical creation processes, and teaching shoppers about the ecological outcomes of their decisions.

7. **Ecological Instruction and Promotion**

Building a practical future requires broad ecological proficiency. Schooling and promotion endeavors assume a vital part in bringing issues to light about ecological issues, cultivating a feeling of obligation, and empowering people, networks, and legislatures to make a significant move

**Worldwide Drives for Natural Supportability**

**Peaceful accords and Arrangements**

Coordinated effort on a worldwide scale is fundamental for tending to ecological difficulties. Peaceful accords, for example, the Paris Settlement on environmental change, the Show on Natural Variety, and the Economical Improvement Objectives (SDGs) laid out by the Assembled Countries, give structures to nations to cooperate towards normal ecological goals.

**Corporate Maintainability Practices**

Organizations and businesses assume a crucial part in ecological maintainability. Many organizations are embracing feasible practices, like lessening fossil fuel byproducts, carrying out eco-accommodating assembling processes, and integrating social obligation into their plans of action.

**Local area Drove Drives**

Nearby people group frequently lead manageability drives custom-made to their particular necessities and difficulties. Local area gardens, squander decrease programs, and environmentally friendly power projects at the grassroots level add to a granular perspective to ecological supportability.

## Mechanical Advancements for Ecological Supportability

### Green Advances

Headways in innovation offer imaginative answers for ecological difficulties. Green advances, including energy-productive apparatuses, electric vehicles, and shrewd matrix frameworks, add to diminishing the environmental impression of human exercises.

### Remote Detecting and Observing

Satellite innovation and remote detecting devices empower researchers to screen changes in the climate, track deforestation, evaluate the soundness of biological systems, and accumulate significant information for informed dynamic in ecological administration.

### Biotechnology for Protection

Biotechnology assumes a part in protection endeavors by offering devices for territory rebuilding, species protection, and hereditary variety conservation. Strategies, for example, quality altering might add to the versatility of biological systems and species confronting ecological dangers.

## Difficulties and Reactions of Ecological Maintainability Endeavors

### Intricacy and Interconnectedness

The intricacy and interconnected nature of ecological frameworks present difficulties to planning and executing successful maintainability measures. Potentially negative side-effects and criticism circles might confound endeavors to resolve explicit issues disregarding more extensive environmental elements.

### Monetary Contemplations

Offsetting monetary interests with ecological maintainability is a tenacious test. Pundits contend that tough natural guidelines might upset financial development, especially in businesses that depend vigorously on asset extraction and creation processes with huge ecological effect.

### Worldwide Disparity and Natural Equity

Natural debasement frequently lopsidedly influences minimized networks and emerging nations. Accomplishing natural equity requires resolving issues of value, guaranteeing that the weight of ecological damage and the advantages of supportability are dispersed decently.

### Momentary Center versus Long haul Manageability

In a world frequently determined by transient financial and political contemplations, there can be a distinction between quick needs and the drawn out objectives of natural manageability. Finding some kind of harmony that tends to dire necessities while guaranteeing the prosperity of people in the future is a tenacious test.

### 6.1 Environmental impact of sodium mineral extraction

Sodium is a urgent component with boundless applications in different enterprises, going from the creation of synthetic compounds and drugs to its broad use in sodium-particle batteries. While sodium extraction is fundamental for satisfying the developing worldwide need, it is similarly essential to survey and address the ecological effect related with its extraction processes. This thorough examination digs into the natural ramifications of sodium mineral extraction, looking at the key cycles, difficulties, and possible answers for limiting the environmental impression.

**Sodium and Its Extraction**

1. **Significance of Sodium**

   Sodium, an exceptionally receptive salt metal, assumes an imperative part in various modern and mechanical applications. It is a critical part of normal salt (sodium chloride) and is fundamental for different natural capabilities, remembering nerve signal transmission and osmotic equilibrium for living creatures. The interest for sodium has flooded as of late because of its utilization in synthetic cycles, metallurgy, and the thriving field of energy stockpiling.

2. **Sodium Mineral Sources**

   Sodium is principally separated from minerals like halite (rock salt), trona, and saline solution stores. Halite and trona are the most well-known minerals utilized for sodium extraction, with huge stores found internationally. Saline solution stores, frequently connected with salt pads, are one more wellspring of sodium and are taken advantage of for their high sodium chloride content.

3. **Sodium Extraction Cycles**

1. **Halite Extraction**

   Halite, a translucent mineral made out of sodium chloride, is mined through customary techniques like penetrating and impacting. When separated, the mineral goes through processing and pounding cycles to get sodium chloride, which is then purged through different techniques, including disintegration and recrystallization.

2. **Trona Extraction**

   Trona, a sodium carbonate mineral, is commonly mined in its regular state. The metal is then handled through a progression of steps that include pounding, warming, and substance treatment to yield sodium carbonate. The sodium carbonate acquired can additionally go through handling to deliver sodium bicarbonate or other sodium compounds.

3. **Saline solution Extraction**

Brackish water extraction includes siphoning sodium-rich saline solution from underground repositories, usually tracked down in salt pads. The saline solution is then handled to remove sodium chloride through strategies, for example, sun based dissipation or mechanical vanishing utilizing heat. The separated sodium chloride is additionally handled to acquire high-virtue sodium.

**Ecological Effect of Sodium Mineral Extraction**

1. **Energy Utilization**

   One of the essential ecological worries related with sodium extraction is the significant energy utilization during the extraction and purging cycles. Energy-concentrated techniques like processing, smashing, and warming add to high ozone depleting substance emanations, especially on the off chance that the energy utilized is gotten from non-inexhaustible sources.

2. **Living space Disturbance and Scene Change**

   Mining activities, particularly those extricating halite and trona, can prompt living space interruption and change of scenes. Open-pit mining, a typical technique for sodium mineral extraction, frequently brings about the evacuation of vegetation, soil, and rock layers, prompting territory obliteration for neighborhood verdure.

3. **Water Use and Pollution**

   Both trona and halite extraction processes require significant water use for processing, washing, and dissolving the minerals. In districts where water assets are scant, this can prompt expanded contest for water, affecting nearby environments and networks. Moreover, the release of wastewater from extraction cycles might contain poisons that can taint close by water sources.

4. **Compound Use and Contamination**

   Compound handling steps in sodium extraction, especially in trona handling, include the utilization of different reagents and solvents. The arrival of these synthetic compounds into the climate can prompt soil and water contamination, influencing the nature of both earthly and amphibian biological systems. Tainting of soil and water with synthetic substances utilized in the extraction and cleaning processes represents a danger to biodiversity and human wellbeing.

5. **Ozone depleting substance Outflows**

   The energy-concentrated nature of sodium extraction adds to critical ozone depleting substance discharges, especially in the event that the energy is gotten from petroleum products. Carbon dioxide ($CO_2$) discharges from the burning of petroleum products in mining gear and handling offices add to environmental change, worsening the natural effect of sodium extraction.

6. **Saline solution Extraction and Environment Effect**

Brackish water extraction processes, while less troublesome regarding scene adjustment, can have explicit natural outcomes. The expulsion of saline solution from underground repositories can adjust the hydrological balance, affecting the strength of biological systems in salt pads. Changes in salt water levels might influence the widely varied vegetation adjusted to these one of a kind conditions.

**Challenges in Moderating Natural Effect**

1. **Energy Progress Difficulties**

   Diminishing the natural effect of sodium extraction requires a change to cleaner and more practical energy sources. Nonetheless, the change from petroleum products to environmentally friendly power in the mining and handling businesses is trying because of the great beginning expenses, existing framework, and the irregular idea of a few sustainable sources.

2. **Mechanical Requirements**

   The turn of events and execution of harmless to the ecosystem advances for sodium extraction face mechanical limitations. While there have been headways in cleaner extraction strategies, the versatility and monetary attainability of these advances remain moves that should be tended to.

3. **Water The board**

   Successful water the executives is urgent in moderating the ecological effect of sodium extraction. Carrying out water-proficient advancements and reusing cycles can assist with limiting the utilization of freshwater and diminish the release of sullied wastewater.

4. **Administrative and Consistence Issues**

   In certain locales, careless natural guidelines or deficient authorization can add to ecologically hurtful extraction rehearses. Reinforcing administrative structures and guaranteeing consistence with natural principles are fundamental for moderating the effect of sodium extraction.

5. **Adjusting Financial and Natural Needs**

Finding some kind of harmony between financial needs and natural maintainability is a tenacious test. Financial contemplations, like the interest for sodium and the benefit of extraction activities, frequently overshadow natural worries, prompting unreasonable practices.

**Expected Answers for Supportable Sodium Extraction**

1. **Sustainable power Coordination**

   Incorporating environmentally friendly power sources, for example, sunlight based and wind power, into sodium extraction cycles can fundamentally diminish ozone harming substance emanations. Interests in environmentally friendly power foundation and motivators for the reception of clean energy advancements can speed up the progress to more reasonable extraction rehearses.

2. **Development in Extraction Advancements**

   Innovative work endeavors ought to zero in on advancing extraction advancements that limit energy utilization and ecological effect. Green science draws near, effective processing cycles, and novel extraction strategies can add to more economical sodium extraction.

3. **Water Reusing and Preservation**

Carrying out water reusing and protection measures can mitigate the tension on freshwater assets. Shut circle frameworks and proficient water the board practices can diminish the natural effect of sodium extraction processes, especially in regions where water shortage is a worry.

4. **Ecological Observing and Announcing**

Laying out vigorous ecological observing and announcing components is fundamental for guaranteeing consistence with guidelines and recognizing regions for development. Straightforwardness in revealing natural execution can likewise cultivate responsibility and empower the reception of feasible practices in the sodium extraction industry.

5. **Partner Commitment and Cooperation**

Drawing in with neighborhood networks, ecological associations, and different partners is basic for creating maintainable sodium extraction rehearses. Co-operative endeavors can prompt the distinguishing proof of ecologically touchy regions, the execution of best practices, and the fuse of nearby information into extraction activities.

6. **Examination into Elective Sources and Innovations**

Investigating elective wellsprings of sodium and creative extraction innovations can differentiate the inventory network and diminish reliance on naturally effective strategies. Examination into elective minerals with lower environmental impressions and the advancement of more maintainable extraction processes are fundamental for the drawn out reasonability of sodium extraction.

**6.2 Strategies for mitigating environmental consequences**

Natural outcomes, originating from human exercises and modern cycles, present huge dangers to environments, biodiversity, and generally speaking planetary well-being. Relieving these outcomes requires a complex methodology that envelops strategy intercessions, mechanical developments, manageable practices, and public mindfulness drives. In this far reaching investigation, we dive into key procedures for moderating natural results, perceiving the interconnectedness of biological frameworks and the basic to encourage an amicable connection among mankind and the climate.

1. **Strategy and Administration Mediations**
1. **Reinforcing Natural Guidelines**

One basic procedure for relieving natural results is the foundation and imple-mentation of hearty ecological guidelines. Legislatures and worldwide bodies assume a pivotal part in setting principles for poison emanations, garbage removal, and regular asset the board. Stricter guidelines, when really upheld, go about as an obstacle against naturally destructive practices and urge businesses to take on cleaner and more maintainable techniques.

2. **Execution of Market-Based Instruments**

   Market-based instruments, for example, carbon evaluating and emanations exchanging frameworks, give financial motivations to organizations to diminish their ecological effect. By putting a cost on fossil fuel byproducts or permitting the exchanging of discharge stipends, these components empower the reception of cleaner innovations and practices. They likewise make a monetary inspiration for organizations to develop and put resources into economical arrangements.

3. **Preservation Approaches and Safeguarded Regions**

   Safeguarding normal natural surroundings and laying out safeguarded regions are basic techniques for alleviating the ecological results of living space misfortune, biodiversity decline, and biological system disturbance. Protection strategies ought to zero in on keeping up with and reestablishing natural equilibrium by shielding crucial environments. Appropriately oversaw safeguarded regions add to biodiversity preservation, soil insurance, and the general strength of environments.

4. **Maintainable Asset The board**

Carrying out maintainable asset the executives rehearses is fundamental for moderating the ecological effect of asset extraction. This incorporates embracing economical ranger service works on, managing fishing to forestall overexploitation, and advancing mindful mining techniques. Practical asset the executives plans to offset human necessities with the conservation of biological systems, guaranteeing the drawn out soundness of normal assets.

**II. Mechanical Advancements and Green Arrangements**

1. **Progress to Environmentally friendly power**

   Moving from petroleum derivatives to sustainable power sources is a foundation procedure for relieving ecological outcomes related with environmental change and air contamination. Sun based, wind, hydroelectric, and geothermal energy offer cleaner options, decreasing ozone depleting substance outflows and reducing reliance on limited assets. Interest in innovative work can additionally improve the productivity and moderateness of environmentally friendly power advances.

2. **Green Innovations and Round Economy**

   The reception of green innovations, like energy-productive apparatuses, eco-accommodating assembling cycles, and waste decrease methods, adds to limiting natural results. The round economy model, underlining reusing, reusing, and diminishing waste, helps decline the extraction of unrefined components and reduces the natural weight related with assembling and removal.

3. **Reasonable Farming Practices**
The horticultural area is a huge supporter of natural results, including deforestation, soil corruption, and water contamination. Reasonable agribusiness rehearses, like natural cultivating, agroecology, and accuracy cultivating, plan to decrease the ecological effect of food creation. These practices focus on soil well-being, biodiversity protection, and water preservation, advancing a more feasible and strong farming framework.

4. **Green Framework and Metropolitan Preparation**

Urbanization and framework improvement frequently bring about environment misfortune, air and water contamination, and expanded energy utilization. Green framework and maintainable metropolitan arranging include planning urban communities and foundation to limit ecological effect. This incorporates green rooftops, penetrable asphalts, and the reconciliation of green spaces to upgrade biodiversity, further develop air quality, and establish more feasible living conditions.

**III. Local area Commitment and Instruction**

1. **Natural Training Projects**
Bringing issues to light and advancing natural proficiency are urgent parts of alleviating ecological results. Instructive projects at different levels, from schools to local area associations, assist with cultivating a feeling of obligation and a more profound comprehension of the interconnectedness between human exercises and the climate. Informed people are bound to go with ecologically cognizant decisions in their day to day routines.

2. **Local area Drove Drives**
Engaging nearby networks to take responsibility for issues is a powerful procedure. Local area drove drives, for example, tree establishing efforts, squander tidy up drives, and reasonable cultivating projects, add to natural protection as well as reinforce the feeling of local area obligation. These drives can be custom fitted to address explicit natural difficulties looked by nearby networks.

3. **Maintainable Ways of life and Purchaser Decisions**

Empowering maintainable ways of life and dependable shopper decisions is a strong procedure for moderating ecological results. People can decrease their biological impression by selecting harmless to the ecosystem items, rehearsing energy and water preservation, and supporting organizations that focus on maintainability. The aggregate effect of cognizant buyer decisions can drive positive change across businesses.

**IV. Global Coordinated effort and Associations**

1. **Worldwide Natural Arrangements**
   Worldwide joint effort is fundamental for tending to ecological difficulties that rise above public lines. Worldwide natural arrangements, for example, the Paris Settlement on environmental change and the Show on Organic Variety, give systems to nations to cooperate on shared ecological objectives. Responsibilities to lessen ozone depleting substance discharges, safeguard biodiversity, and advance supportable improvement encourage a brought together way to deal with moderating natural outcomes.

2. **Corporate Social Obligation (CSR) Drives**
   Organizations and enterprises play a critical part in moderating ecological outcomes through dependable practices and reasonable turn of events. Corporate Social Obligation (CSR) drives that focus on ecological stewardship, moral obtaining, and local area commitment add to a more economical and comprehensive worldwide economy. Organizations between organizations, NGOs, and legislatures can drive positive change and development.

3. **Examination and Innovation Move**

Worldwide joint effort in examination and innovation move works with the sharing of information, skill, and reasonable advances. Non-industrial nations, specifically, can profit from the exchange of harmless to the ecosystem advancements that help jump customary, asset escalated improvement pathways. Cooperative exploration endeavors can prompt creative answers for relieving natural results.

**V. Transformation and Flexibility Building**

1. **Environmental Change Transformation Methodologies**
   Given the continuous effects of environmental change, transformation methodologies are pivotal for building flexibility in networks and biological systems. This incorporates measures like the improvement of environment versatile foundation, early admonition frameworks for outrageous climate occasions, and the advancement of environment brilliant agribusiness. Transformation methodologies assist networks with adapting to the changing environment and limit related natural results.

2. **Biological system Reclamation and Restoration**
   Reestablishing and restoring debased biological systems is a proactive methodology for relieving ecological outcomes. Afforestation, reforestation, and wetland reclamation projects add to biodiversity preservation, carbon sequestration, and the upgrade of biological system administrations. Environment rebuilding endeavors can be local area driven or part of bigger public and worldwide reclamation drives.

3. **Fiasco Hazard Decrease (DRR)**

Executing Fiasco Chance Decrease methodologies is fundamental for limiting the ecological outcomes of catastrophic events. This implies surveying and relieving gambles with connected with floods, tropical storms, quakes, and different fiascos. Arranging and foundation advancement that consider potential natural effects can add to decreasing weakness and improving strength even with debacles.

**VI. Checking, Assessment, and Detailing**

1. **Ecological Effect Appraisal (EIA)**
   Orderly assessment of expected ecological outcomes prior to starting tasks is basic. Natural Effect Appraisals (EIA) help recognize and relieve potential ecological dangers related with advancement projects. States and ventures ought to focus on the consolidation of EIAs into dynamic cycles to guarantee that natural results are satisfactorily tended to.

2. **Straightforward Announcing and Responsibility**
   Straightforward announcing of ecological execution by states, organizations, and associations is fundamental for responsibility and improvement. Openly available ecological reports that detail moves initiated, progress made, and regions requiring consideration empower partners to consider substances responsible. This straightforwardness cultivates a culture of liability and energizes constant improvement in ecological administration.

3. **Incorporated Observing Frameworks**

Creating coordinated observing frameworks that track different ecological pointers, for example, air and water quality, biodiversity levels, and environment boundaries, gives fundamental information to independent direction. These frameworks empower brief reactions to arising ecological issues and work with versatile administration draws near. Incorporation guarantees a comprehensive comprehension of ecological results across various areas.

**6.3 Regulatory frameworks and industry responsibility**

Administrative structures and industry obligation assume vital parts in molding the ecological, social, and moral scene of different areas. The transaction between unofficial laws and corporate obligation shapes a nexus that decides the general effect of businesses in the world and society. In this far reaching investigation, we dive into the meaning of administrative systems, the development of industry obligation, and how the joint effort between the two can drive reasonable practices, encourage advancement, and guarantee a harmony between financial development and natural stewardship.

1. **The Development of Administrative Systems**
1. **Verifiable Setting**
   Administrative systems, generally established in tending to general wellbeing and

security concerns, have developed to envelop a more extensive scope of issues, including ecological insurance, work privileges, and corporate administration. The modern transformation in the nineteenth century featured the requirement for guidelines to relieve the adverse consequences of fast industrialization, like contamination, risky working circumstances, and abuse of regular assets.

2. **Natural Guidelines**

   The last 50% of the twentieth century saw a huge extension of natural guidelines worldwide. Occasions like the Cuyahoga Stream bursting into flames in 1969 and the distribution of Rachel Carson's "Quiet Spring" featured the critical requirement for natural securities.

   Administrative structures arose to address air and water contamination, risky garbage removal, and the preservation of regular territories. Milestone regulations, including the Spotless Air Act and the Perfect Water Act in the US, set up for thorough ecological guideline.

3. **Social and Moral Guidelines**

Notwithstanding ecological worries, administrative structures have extended to address social and moral contemplations. Work privileges, work environment well-being, and moral strategic approaches are presently essential parts of administrative systems. Worldwide bodies, like the Worldwide Work Association (ILO), have created norms to guarantee fair work rehearses universally, building up the possibility that organizations should work morally and dependably.

**II. Industry Obligation: Corporate Social Obligation (CSR) and Then some**

1. **Rise of Corporate Social Obligation (CSR)**

   As cultural assumptions for organizations developed, another worldview known as Corporate Social Obligation (CSR) arose. CSR goes past simple consistence with guidelines and urges organizations to proactively add to cultural prosperity and ecological manageability. The idea acquired unmistakable quality in the late twentieth hundred years, and many organizations started integrating CSR drives into their plans of action.

2. **Triple Primary concern: Individuals, Planet, Benefit**

   The idea of the Triple Primary concern (TBL) typifies that organizations ought to be responsible for monetary execution (benefit) yet in addition for their effect on individuals (social) and the planet (ecological). This all encompassing way to deal with business stresses the interconnectedness of monetary, social, and natural factors and highlights the job of organizations in contributing emphatically to society.

3. **Supportable Improvement Objectives (SDGs)**

The Unified Countries' Supportable Improvement Objectives (SDGs) give a complete system to tending to worldwide difficulties, including neediness, imbalance, environmental change, ecological corruption, harmony, and equity. Numerous organizations have adjusted their CSR endeavors to the SDGs, perceiving the significance of adding to an additional supportable and evenhanded world.

## III. The Nexus: Administrative Consistence and Willful Responsibilities

1. **Cooperative energy Among Guidelines and CSR**

   The nexus between administrative structures and industry obligation lies in the cooperative energy between required consistence and deliberate responsibilities. While guidelines set gauge norms and lawful commitments for organizations, CSR drives permit organizations to go past these necessities and contribute decidedly to society and the climate. The blend of both makes a more vigorous and far reaching way to deal with manageability.

2. **Administrative Consistence as a Beginning stage**

   Compulsory consistence with guidelines fills in as the beginning stage for dependable business lead. Complying with natural, work, and moral guidelines is non-debatable and lays out the standard for moral business tasks. Administrative structures set the standards of the game, characterizing what is legitimately reasonable and giving a level battleground to organizations.

3. **CSR as an Impetus for Development**

Past consistence, industry obligation, as appeared through CSR drives, goes about as an impetus for development. Organizations that embrace CSR frequently end up at the front line of advancement, creating maintainable practices, eco-accommodating items, and socially mindful plans of action. CSR turns into an essential device for organizations to separate themselves on the lookout, draw in upright customers, and improve their drawn out suitability.

## IV. Key Parts of Administrative Systems

1. **Ecological Guidelines**

   Ecological guidelines center around forestalling and alleviating the effect of modern exercises on biological systems, air, water, and soil. These guidelines set discharge norms, garbage removal rules, and preservation prerequisites to protect regular assets and biodiversity. Models incorporate as far as possible for poisons, limitations on perilous garbage removal, and safeguarded regions regulation.

2. **Work Guidelines**

   Work guidelines are intended to safeguard the freedoms and prosperity of laborers. They cover viewpoints like the lowest pay permitted by law, working hours, word related wellbeing and security, and opportunity of affiliation.

These guidelines guarantee fair treatment, safe working circumstances, and evenhanded remuneration for representatives. The ILO shows give a worldwide system to work guidelines that states and organizations are urged to take on.

3. **Moral Strategic policies**

Administrative systems additionally address moral contemplations in business tasks. Hostile to defilement regulations, fair rivalry guidelines, and straightforwardness prerequisites expect to establish a business climate based on respectability and moral direct. These guidelines encourage trust among partners and add to a level battleground where organizations contend in view of legitimacy and moral way of behaving.

**V. Industry Obligation in real life**

1. **Economical Stockpile Chains**
   Organizations are progressively perceiving the significance of economical stockpile chains. Mindful obtaining of natural substances, fair treatment of providers, and adherence to moral work rehearses all through the inventory network are indispensable parts of industry obligation. This includes directing expected level of effort to guarantee that providers maintain ecological and social guidelines.

2. **Environment Activity and Carbon Impartiality**
   Many organizations are going to proactive lengths to address environmental change by focusing on carbon impartiality and lessening their carbon impression. This frequently includes setting emanation decrease targets, putting resources into sustainable power, and carrying out energy-effective practices. Industry obligation with regards to environment activity lines up with worldwide endeavors to restrict an unnatural weather change and progress to a low-carbon economy.

3. **Social Effect Drives**

CSR drives frequently reach out to social effect programs that address squeezing cultural issues. This can remember speculations for instruction, medical care, destitution mitigation, and local area improvement. By effectively captivating in drives that add to the prosperity of networks, organizations satisfy their job as dependable corporate residents.

**VI. Difficulties and Valuable open doors**

1. **Challenges in Administrative Consistence**
   Challenges in administrative consistence can emerge because of changing guidelines across purviews, lacking authorization components, or the intricacy of developing guidelines. Organizations working in different areas should explore assorted administrative scenes, making consistence testing. Also, more modest

organizations might confront monetary imperatives in gathering rigid administrative necessities.

2. **Adjusting Monetary Development and Ecological Stewardship**
   A perpetual test is finding some kind of harmony between monetary development and ecological stewardship. Some contend that rigid ecological guidelines might frustrate financial turn of events, especially in ventures with high natural effect. Accomplishing an agreeable equilibrium requires a nuanced approach that supports maintainable practices without smothering financial advancement.

3. **Straightforwardness and Responsibility**
   Guaranteeing straightforwardness and responsibility in both administrative consistence and CSR drives is fundamental. Absence of straightforwardness can disintegrate trust in organizations, making it basic for organizations to give clear data about their natural and social execution. Outsider accreditations and free reviews add to responsibility and believability.

4. **Embracing Advancement**

Embracing development is both a test and an open door. Organizations that view administrative consistence and industry obligation as any open doors for advancement are better situated for long haul achievement. Manageable plans of action, green advances, and round economy rehearses address roads for development that line up with ecological and social goals.

**VII. Worldwide Coordinated efforts and Future Bearings**

1. **Worldwide Joint effort for Economical Turn of events**
   Worldwide difficulties require worldwide arrangements, and global coordinated efforts are pivotal for resolving natural and social issues. State run administrations, organizations, and common society should cooperate to create and carry out manageable practices. Drives like the Unified Countries Worldwide Smaller and cooperative endeavors, for example, the Competition to Zero mission show the significance of an assembled front in handling worldwide difficulties.

2. **Mix of ESG Standards**
   The mix of Ecological, Social, and Administration (ESG) standards in strategic approaches is picking up speed. Financial backers and purchasers progressively consider ESG factors while simply deciding, impacting organizations to take on additional dependable and supportable practices. ESG joining includes evaluating an organization's effect on the climate, its social obligation, and the viability of its administration structures.

3. **Proceeded with Development of Administrative Structures**

Administrative structures will keep on advancing to address arising difficulties and open doors. As logical comprehension progresses and new natural and social issues come to the very front, guidelines should adjust to relieve gambles really. Adaptability in administrative systems considers responsiveness to changing conditions while keeping a guarantee to maintainability.

# Chapter 7

## Future Prospects And Innovations

What's in store holds monstrous commitment and potential for creative arrangements that can address squeezing worldwide difficulties, going from ecological maintainability and medical care to mechanical headways that rethink how we live and function. As we stand on the slope of another time, a few critical areas of center are arising, driving the direction of future possibilities and developments.

1. **Feasible Advances and Green Developments**
1. **Environmentally friendly power Progressions**
   The continuous quest for environmentally friendly power sources is ready to alter the worldwide energy scene. Advancements in sun powered and wind innovations, combined with forward leaps in energy capacity, vow to make clean energy more available, proficient, and practical. Progressions in cutting edge sunlight based cells, shrewd frameworks, and energy stockpiling arrangements are significant for accomplishing an economical and low-carbon future.
2. **Roundabout Economy Practices**

The progress towards a round economy, where items are intended for reuse, reusing, and insignificant natural effect, is picking up speed. Developments in material science, squander the executives, and item configuration are fundamental for understanding a round economy. From biodegradable bundling to shut circle fabricating processes, these advancements intend to diminish squander and advance the practical utilization of assets.

2. **Medical care and Biotechnological Forward leaps**

1. **Accuracy Medication and Customized Treatments**
   The eventual fate of medical services is progressively customized, with headways in genomics and information examination preparing for accuracy medication.

Fitting clinical medicines to a person's hereditary cosmetics takes into consideration more viable and designated treatments, working on persistent results and diminishing secondary effects.

The coordination of man-made brainpower (computer based intelligence) further speeds up the examination of immense datasets, working with fast progressions in clinical exploration and finding.

2. **Biotechnology and Quality Altering**

The area of biotechnology, especially quality altering advancements like CRISPR-Cas9, holds groundbreaking potential for treating hereditary issues and upgrading rural practices. As these innovations develop, moral contemplations and dependable administration will assume a vital part in directing their applications. Developments in quality altering might change sickness counteraction, crop versatility, and environment rebuilding.

**3. Innovative Changes: artificial intelligence, Quantum Figuring, and Then some**

1. **Man-made reasoning (artificial intelligence) and AI**
   Man-made intelligence and AI are set to rethink enterprises via computerizing processes, improving direction, and empowering new degrees of productivity. From prescient examination in medical care to independent vehicles and brilliant urban communities, man-made intelligence applications are different and far reaching. Continuous exploration centers around tending to moral worries, inclination relief, and the dependable sending of computer based intelligence advances.

2. **Quantum Figuring**

Quantum figuring addresses a change in perspective in computational power, possibly tackling complex issues at speeds impossible with old style PCs. Quantum PCs can possibly upset fields like cryptography, enhancement, and medication disclosure. Nonetheless, huge specialized difficulties remain, and the acknowledgment of useful quantum processing applications is still in the beginning phases of improvement.

**4. Network and 5G Innovation**

1. **5G Organizations and Web of Things (IoT)**
   The rollout of 5G organizations is set to change network, empowering quicker information transmission and lower inactivity. This is basic for the expansion of IoT gadgets, making a flawlessly associated world. From savvy homes and urban areas to modern applications, the blend of 5G and IoT opens roads for remarkable development and proficiency.

2. **Edge Processing**

Edge figuring, a decentralized registering model, supplements the capacities of 5G by handling information nearer to the source instead of depending on incorporated cloud servers.

This decreases dormancy as well as upgrades protection and security. Edge registering is essential to the proficient working of IoT gadgets and is ready to reform how information is handled and used.

### 5. Environment Strength and Natural Checking

1. **Environment Advancements**
   Notwithstanding environmental change, inventive advances are essential for checking, adjusting to, and moderating natural difficulties. Remote detecting, satellite advancements, and environment demonstrating are propelling comprehension we might interpret environment designs. Environment strength innovations, including progressed climate expectation models and feasible rural practices, add to building a stronger world.

2. **Natural Sensors and Observing Frameworks**

The multiplication of ecological sensors and checking frameworks gives constant information on air quality, water levels, and biodiversity. These advances enable networks, state run administrations, and businesses to go with informed choices for reasonable asset the executives. Coordinating information driven bits of knowledge into ecological strategies is fundamental for moderating the effect of human exercises on biological systems.

### 6. Schooling and Labor force Developments

1. **Deep rooted Learning Stages**
   The fate of work is developing quickly, determined by mechanical headways and changing monetary scenes. Deep rooted learning stages controlled by artificial intelligence and computer generated reality are arising, permitting people to consistently upskill and adjust to advancing position necessities. These stages offer customized growth opportunities and add to a more lithe and strong labor force.

2. **Remote Work Advances**

The worldwide shift towards remote work, advanced by the Coronavirus pandemic, has prodded developments in coordinated effort apparatuses, virtual correspondence, and computerized efficiency arrangements. Future working environment innovations are probably going to focus on adaptability, inclusivity, and representative prosperity, reshaping conventional ideas of work and encouraging a more associated and versatile labor force.

### 7. Moral Tech and Mindful Development

1. **Moral Plan Standards**

   As innovation turns out to be more imbued in day to day existence, there is a developing accentuation on moral plan standards. Dependable tech development includes thinking about the social and moral ramifications of items and administrations from their initiation. This incorporates resolving issues of security, predisposition in calculations, and the cultural effect of arising advances.

2. **Corporate Social Obligation (CSR) in Tech**

Tech organizations are progressively perceiving their cultural effect and the requirement for capable strategic approaches. Coordinating CSR into the tech business includes thinking about the ecological impression of gadgets, advancing computerized incorporation, and resolving social issues exacerbated by innovation. This shift toward moral tech rehearses lines up with buyer assumptions for socially cognizant organizations.

**8. Worldwide Joint efforts for Reasonable Turn of events**

1. **Global Associations**

   Tending to worldwide difficulties requires cooperative endeavors on a global scale. Multilateral associations between legislatures, confidential area substances, and non-administrative associations are urgent for accomplishing manageable improvement objectives. Drives that cultivate information sharing, innovation move, and facilitated activity add to a more interconnected and versatile worldwide local area.

2. **Open Source Advancement**

Open-source advancement, where innovations and arrangements are shared unreservedly, encourages coordinated effort and speeds up progress. This cooperative model has been especially significant in fields like programming advancement, and its standards are reaching out to regions like medical services, environmentally friendly power, and supportable horticulture. Open-source drives add to democratizing advancement and defeating obstructions to advance.

**7.1 Emerging technologies in sodium mineral utilization**

Sodium, a flexible salt metal, is seeing a flood popular across different enterprises, moving innovative work endeavors toward inventive advances for its extraction, handling, and usage. This investigation dives into arising advances in sodium mineral usage, featuring progressions that can possibly reform the sodium business, upgrade manageability, and open new applications for this fundamental component.

1. **Economical Sodium Extraction Innovations**
1. **Electrochemical Extraction**

   Conventional strategies for sodium extraction, for example, the Downs cycle, are energy-serious and add to ecological effect. Arising electrochemical extraction innovations offer a more supportable other option. These techniques include the utilization of electrolysis to separate sodium from its mineral sources. By utilizing headways in terminal materials and energy-effective cycles, electrochemical extraction lessens both energy utilization and natural impression.

2. **Green Dissolvable Extraction**

The utilization of harmless to the ecosystem solvents in sodium extraction processes is building up some forward movement. Conventional techniques frequently include the utilization of solvents with ecological disadvantages. Green dissolvable extraction uses eco-accommodating choices that are biodegradable and reducedly affect biological systems. This development lines up with the developing accentuation on supportable practices in the mining and mineral extraction industry.

**2. Sodium Reusing and Reuse Innovations**

1. **Sodium-Particle Batteries**
   As the interest for energy capacity arrangements develops, sodium-particle batteries have arisen as a promising option in contrast to lithium-particle batteries. Sodium, being more plentiful and practical than lithium, makes these batteries an alluring choice. Research is centered around growing elite execution cathode and anode materials, as well as improving the electrolyte creation, to upgrade the effectiveness and life span of sodium-particle batteries. These developments could upset the energy stockpiling scene, particularly in enormous scope applications.

2. **Sodium Recuperation from Squander Streams**

Endeavors are in progress to foster advances that empower the recuperation of sodium from different waste streams, including modern effluents and brackish water arrangements. By executing effective partition and recuperation processes, enterprises can limit sodium squander, decrease ecological effect, and make a shut circle framework. This approach lines up with the standards of round economy and asset proficiency.

**3. High level Handling and Refining Innovations**

1. **Liquid Salt Advancements**
   Liquid salt advancements are being investigated for the handling and refining of sodium. These advances influence the interesting properties of liquid salts to work with high-temperature processes with upgraded effectiveness.
   Liquid salt electrolysis, for example, is a technique being scrutinized for sodium extraction and refining. The utilization of liquid salts can add to lessening energy utilization in sodium handling.

## 2. High level Filtration and Partition Methods

In the refining and purging of sodium, high level filtration and detachment strategies are acquiring significance. Layer advances, for example, film refining and particle trade films, are being utilized to accomplish high-virtue sodium. These strategies offer exact command over the partition cycle and add to limiting waste and energy utilization during sodium refinement.

## 4. Nanotechnology Applications in Sodium Usage

### 1. Nanostructured Sodium Materials

Nanotechnology is assuming an extraordinary part in the improvement of nanostructured sodium materials with upgraded properties. Nanostructured cathodes for sodium-particle batteries, for example, display further developed execution concerning energy thickness and cycling solidness. The use of nanotechnology in sodium usage reaches out to different businesses, including hardware, where nanostructured materials empower the making of more modest and more productive gadgets.

### 2. Nanofluids in Intensity Move Applications

In modern cycles including the utilization of sodium, for example, heat move frameworks in atomic reactors and concentrated sun oriented power plants, nanofluids are being investigated for their capability to improve heat move productivity. The expansion of nanoparticles to sodium-based heat move liquids can work on warm conductivity, prompting more productive energy move and generally framework execution.

## 5. Sodium in Cutting edge Materials and Assembling

### 1. Sodium-Based Impetuses

Sodium-based impetuses are acquiring consideration in synthetic cycles and modern assembling. These impetuses offer benefits, for example, minimal expense and high reactivity, making them reasonable for different synergist responses. Research is centered around enhancing the properties of sodium-based impetuses for applications in natural union, petrochemical refining, and ecological remediation.

### 2. Sodium in Cutting edge Combinations

The utilization of sodium in cutting edge composites is growing, driven by endeavors to improve the mechanical and warm properties of materials. Sodium-containing compounds show special attributes, including further developed strength and erosion opposition. These combinations track down applications in aviation, car, and underlying designing, adding to the advancement of lightweight and strong materials.

### 6. Natural and Wellbeing Advancements in Sodium Taking care of

1. **Fluid Sodium Coolants in Cutting edge Reactors**
   Fluid sodium is used as a coolant in specific high level atomic reactor plans. The high warm conductivity and low neutron assimilation properties of fluid sodium go with it a productive decision for moving intensity in atomic power age. Continuous examination plans to upgrade the security and unwavering quality of fluid sodium coolant frameworks, tending to difficulties related with its reactivity with air and water.

2. **Sodium Fire Concealment Innovations**

Given the reactivity of sodium with dampness and air, security estimates in sodium-taking care of offices are of fundamental significance. Arising advances in sodium fire concealment center around quick and compelling strategies to control sodium fires. Developments incorporate the improvement of fire suppressants custom fitted for sodium fires, robotized fire concealment frameworks, and high level fire location advances.

### 7. Difficulties and Contemplations in Sodium Mineral Usage

1. **Security and Reactivity**
   The reactivity of sodium with air and dampness presents security challenges in taking care of and handling. Advancements in control frameworks, idle climates, and wellbeing conventions are fundamental to relieve the dangers related with sodium use. Furthermore, progressions in materials designing expect to make more hearty and consumption safe materials for sodium-taking care of framework.

2. **Natural Effect of Extraction**

While manageable extraction innovations are arising, the ecological effect of enormous scope sodium extraction stays a thought. Limiting territory disturbance, water use, and energy utilization in extraction processes is pivotal for guaranteeing the general manageability of sodium use. Ceaseless investigation into eco-accommodating extraction techniques is basic for tending to these natural worries.

### 8. Future Mix and Collaborations

1. **Mix of Advances**
   The eventual fate of sodium mineral usage lies in the reconciliation of these arising advances into exhaustive, practical frameworks. Consolidating progressed extraction techniques with productive reusing, using sodium in different

applications, and coordinating security measures into sodium-taking care of cycles will add to a more all encompassing and practical sodium industry.

2. **Cooperative energies with Other Arising Advancements**

Investigating cooperative energies with other arising innovations, like man-made reasoning (computer based intelligence) for process advancement and mechanical technology for more secure taking care of, can additionally improve the productivity and wellbeing of sodium usage. Cooperative exploration and interdisciplinary methodologies will be instrumental in opening the maximum capacity of sodium in different modern applications.

### 7.2 Advancements in extraction and processing methods

Progressions in extraction and handling strategies are at the front of changing ventures, empowering more productive and manageable use of assets. From customary techniques that have controlled human civilizations for a really long time to state of the art innovations that influence development and manageability, this investigation dives into the advancement of extraction and handling across different areas.

1. **Advancement of Extraction Strategies**
1. **Conventional Extraction Methods**

    Mankind's set of experiences is interwoven with the advancement of extraction procedures to get fundamental assets. Conventional techniques, like manual mining, agribusiness, and basic refining, established the groundwork for development. These procedures were in many cases work concentrated and restricted in scale, mirroring the mechanical abilities of their separate periods.

2. **Automation and Industrialization**

The coming of the Modern Unrest denoted a critical change in extraction techniques. Motorization, fueled by steam motors and later by power, empowered enormous scope mining, assembling, and handling. This time saw the far reaching utilization of hardware in coal mining, metal extraction, and the development of materials, altogether expanding proficiency and result.

2. **Mechanical Developments in Extraction**

1. **Water powered Breaking (Deep oil drilling) in Oil and Gas Extraction**

    In the twentieth hundred years, the oil and gas industry saw a groundbreaking development with the far and wide reception of pressure driven cracking, usually known as deep earth drilling. This method includes infusing high-pressure liquid into underground stone developments to deliver oil and flammable gas. While deep oil drilling has opened tremendous stores of hydrocarbons, it has likewise raised worries about natural effects, including water defilement and seismic action.

## 2. In Situ Filtering in Mining

In situ filtering, otherwise called arrangement mining, addresses a huge progression in the extraction of minerals, especially uranium. This strategy includes infusing a filtering arrangement into metal stores, permitting the objective minerals to break down and be siphoned to the surface for handling. In situ filtering limits surface aggravation and lessens natural effect contrasted with customary mining strategies.

## 3. Manageable Extraction Advancements

### 1. Green Science in Metal Extraction

Green science standards are progressively forming extraction techniques, especially in metal handling. Advancements center around lessening the natural impression of metal extraction by limiting waste, utilizing harmless to the ecosystem solvents, and enhancing energy proficiency. The joining of green science standards lines up with the more extensive objective of accomplishing manageable and dependable asset usage.

### 2. Biomimicry in Extraction Strategies

Biomimicry includes attracting motivation from nature to foster inventive advancements. In extraction techniques, biomimicry may include recreating regular cycles saw in plants or microorganisms to extricate wanted compounds specifically. This approach can possibly lessen the utilization of cruel synthetic compounds and energy-escalated processes, making more manageable extraction rehearses.

## 4. Leap forwards in Mineral Handling

### 1. High level Arranging Innovations

Current mineral handling includes progressed arranging advances that empower the effective detachment of important minerals from metal. Arranging strategies, including sensor-based arranging and man-made consciousness (artificial intelligence)- driven arranging frameworks, upgrade the accuracy and speed of mineral handling. These innovations further develop asset recuperation rates and lessen the natural effect of mining tasks.

### 2. Microwave Innovation in Mineral Handling

Microwave innovation is arising as a promising device in mineral handling. It includes the utilization of microwaves to specifically warm minerals, working with their division from the metal. This strategy can possibly diminish energy utilization and improve the productivity of mineral handling, making it an alluring choice for the mining business.

## 5. Advancements in Oil and Gas Handling

1. **Carbon Catch and Capacity (CCS)**

   Because of the natural difficulties related with petroleum product extraction and ignition, carbon catch and capacity (CCS) has arisen as a critical development in the oil and gas industry. CCS advancements catch carbon dioxide discharges from modern cycles and power plants, keeping them from entering the air. This approach adds to moderating environmental change while permitting proceeded with utilization of petroleum derivatives.

2. **Improved Oil Recuperation (EOR)**

Improved Oil Recuperation procedures intend to amplify oil extraction from repositories. Developments in EOR incorporate the infusion of synthetic compounds, steam, or gases into oil wells to adjust the properties of the repository and further develop oil recuperation. These techniques assist with expanding the useful existence of oil fields and improve generally speaking extraction effectiveness.

**6. Feasible Farming Practices**

1. **Accuracy Horticulture**

   In the horticultural area, accuracy farming addresses a huge progression in asset usage. Advancements, for example, GPS-directed work vehicles, sensors, and robots empower ranchers to enhance the utilization of water, manures, and pesticides. Accuracy horticulture advances maintainable cultivating rehearses by decreasing waste, limiting natural effect, and further developing in general harvest yield.

2. **Hydroponics and Aquaculture**

Hydroponics and aquaculture are inventive techniques for developing harvests without customary soil-based farming. Hydroponics joins hydroponics (fish cultivating) with tank-farming (soilless plant development), making a cooperative framework where fish squander gives supplements to endlessly establishes help channel and purge water. These techniques preserve water, dispense with soil-related sicknesses, and proposition proficient supplement conveyance to plants.

**7. Mechanical Progressions in Food Handling**

1. **High-Tension Handling (HPP)**

   High-Strain Handling is a non-warm food conservation strategy that utilizations raised tensions to expand the timeframe of realistic usability of food items. HPP holds the nourishing quality and newness of food varieties while dispensing with hurtful microorganisms. This innovation is generally utilized in the handling of juices, meats, and prepared to-eat feasts, giving purchasers negligibly handled, safe, and nutritious food choices.

2. **Nanotechnology in Food Handling**

Nanotechnology is making advances into the food business, offering creative answers for food handling and bundling. Nano-sized particles and materials can be utilized to upgrade the time span of usability of items, further develop sanitation, and empower controlled arrival of supplements. While promising, the utilization of nanotechnology in food handling brings up issues about wellbeing and administrative contemplations.

### 8. Future Possibilities: Industry 4.0 and Then some

1. **Industry 4.0 in Assembling**
   Industry 4.0, described by the reconciliation of computerized innovations into assembling processes, is forming the fate of extraction and handling across ventures. Savvy sensors, information investigation, and network empower constant observing and enhancement of extraction and handling activities. This computerized change upgrades productivity, lessens asset squander, and adds to practical and shrewd asset use.

2. **Independent Frameworks in Mining**

The sending of independent frameworks, including independent vehicles and boring hardware, is upsetting the mining business. These frameworks influence man-made intelligence, AI, and high level sensors to work with insignificant human intercession, further developing security and effectiveness. Independent advancements empower constant activity, diminish personal time, and enhance asset extraction processes.

### 9. Difficulties and Contemplations in Cutting edge Extraction and Handling

1. **Ecological Effect**
   Notwithstanding progressions, the extraction and handling ventures keep on confronting difficulties connected with ecological effect. Huge scope mining, utilization of synthetic substances, and energy-escalated processes add to environment interruption, water contamination, and fossil fuel byproducts. Supportable practices and capable asset the board are basic to relieve these natural difficulties.

2. **Innovative Availability and Moderateness**

While trend setting innovations offer extraordinary arrangements, their inescapable reception might be impeded by issues of openness and reasonableness. Limited scope enterprises and creating areas might confront difficulties in embracing best in class advances because of significant expenses, framework impediments, and the requirement for particular abilities. Connecting this mechanical gap is significant for comprehensive and supportable turn of events.

### 10. Moral Contemplations and Local area Commitment

1. **Social Effects of Extraction**

   Extraction exercises frequently have significant social effects, including uprooting of networks, loss of jobs, and changes in nearby environments. Moral contemplations in extraction and handling include drawing in with nearby networks, regarding native freedoms, and guaranteeing fair pay for those impacted by asset extraction. Straightforward and comprehensive dynamic cycles are fundamental for tending to these moral contemplations.

2. **Capable Stock Chains**

The idea of capable stock chains underlines moral obtaining and creation rehearses. From struggle minerals to practical farming, guaranteeing that assets are removed, handled, and used mindfully all through the inventory network is urgent. Confirmation programs, recognizability frameworks, and partner commitment add to building mindful stock chains that focus on natural and social supportability.

**7.3 Potential role of sodium minerals in a sustainable future**

Sodium, an omnipresent salt metal, assumes an imperative part in different businesses and regular day to day existence. As the world countenances expanding difficulties connected with asset accessibility, ecological supportability, and the change to cleaner energy sources, the expected job of sodium minerals becomes significant in forming an economical future. This investigation digs into the multi-layered commitments of sodium minerals across different areas, featuring their importance in cultivating maintainability and advancement.

1. **Sodium Minerals: Overflow and Availability**

   Sodium is an exceptionally plentiful component, comprising around 2.6% of the World's covering. Its minerals, like halite (rock salt) and trona, are far reaching and effectively open. The overflow of sodium minerals makes them a significant asset for different applications without the international intricacies frequently connected with different components.

2. **Sodium in Clean Energy Advancements**

1. **Sodium-Particle Batteries**

   One of the most encouraging regions where sodium minerals can upset the scene is in energy capacity. Sodium-particle batteries (SIBs) are acquiring consideration as an option in contrast to lithium-particle batteries. Sodium is more plentiful and practical than lithium, making SIBs an appealing choice for enormous scope energy capacity. Research is centered around growing superior execution sodium-particle battery advancements that can rival conventional lithium-particle batteries concerning energy thickness and cycling soundness.

2. **Sodium as an Intensity Move Medium**

Sodium's one of a kind properties, like its high warm conductivity, low consistency, and low neutron retention, make it a brilliant intensity move medium. Fluid sodium is utilized in cutting edge atomic reactors as a coolant, working with the proficient exchange of intensity produced during atomic parting. This application adds to the improvement of economical and effective atomic power age.

**3. Sodium in Feasible Agribusiness**

1. **Soil Improvement and Supplement The board**
   Sodium minerals, when utilized reasonably, can add to supportable farming practices. Sodium can be utilized to further develop soil design and help in supplement the executives. The expansion of sodium-containing mixtures can upgrade soil porousness and water maintenance, especially in clayey soils. Be that as it may, cautious thought and observing are fundamental to forestall antagonistic impacts on soil quality.
2. **Hydroponics and Fish Cultivating**

Sodium carbonate (soft drink debris) is a sodium mineral with applications in hydroponics. It is utilized to control pH levels in water, giving a reasonable climate to fish and other sea-going life forms. Reasonable fish cultivating rehearses frequently include the cautious administration of water quality, and sodium minerals assume a part in keeping up with ideal circumstances for hydroponics.

**4. Sodium in Modern Cycles**

1. **Substance Assembling**
   Sodium compounds are basic in different substance producing processes. Sodium hydroxide (scathing pop) is a vital modern synthetic utilized in the development of materials, paper, and cleansers. The maintainable use of sodium in these cycles includes streamlining creation techniques, limiting waste, and investigating green science standards.
2. **Metallurgical Applications**

In metallurgy, sodium assumes a part in different applications. Sodium can be utilized as a lessening specialist in the extraction of specific metals, adding to additional energy-effective and harmless to the ecosystem processes. The improvement of inventive metallurgical procedures using sodium minerals lines up with the more extensive objective of practical and dependable asset extraction.

**5. Sodium in Water Treatment and Desalination**

1. **Desalination Innovations**
   The worldwide interest for freshwater is heightening, and desalination advancements are turning out to be progressively significant. Sodium minerals,

especially sodium chloride, are bountiful in seawater. While desalination generally depends on eliminating sodium particles alongside different salts, progressions in desalination advances are investigating the potential for specific sodium extraction, adding to more energy-productive desalination processes.

2. **Water Relaxing**

Sodium assumes a part in water relaxing cycles, where hard water, wealthy in calcium and magnesium particles, is treated with sodium particles to forestall the development of scale in lines and machines. Feasible water mellowing rehearses include the mindful utilization of sodium-based water conditioners and investigating elective advancements that limit ecological effect.

### 6. Difficulties and Contemplations in Sodium Mineral Usage

1. **Ecological Effect of Sodium Extraction**
   Enormous scope extraction of sodium minerals, especially from underground stores, can have natural outcomes. Mining exercises might upset biological systems, influence biodiversity, and add to territory debasement. Supportable sodium extraction requires the execution of mindful mining rehearses, living space reclamation endeavors, and the improvement of eco-accommodating extraction innovations.

2. **Sodium in Agribusiness: Difficult exercise**

While sodium can be useful in specific horticultural applications, it is vital for find some kind of harmony to stay away from accidental adverse consequences. Unnecessary sodium in soils can prompt soil saltiness, unfavorably influencing plant development and water quality. Feasible horticultural works on including sodium minerals require cautious checking, soil the board systems, and adherence to suggested application rates.

### 7. Future Developments and Exploration Wildernesses

1. **High level Sodium-Particle Battery Innovations**
   Progressing research in the field of sodium-particle batteries centers around further developing their exhibition attributes to match or outperform lithium-particle batteries. Advancements in cathode materials, electrolyte plan, and assembling processes expect to upgrade energy thickness, cycle life, and by and large proficiency. High level sodium-particle battery innovations hold the possibility to change energy capacity on a worldwide scale.

2. **Economical Extraction Strategies**

The improvement of economical extraction strategies for sodium minerals is a basic examination wilderness. Advancements that lessen energy utilization, limit natural effect, and enhance asset recuperation are fundamental for guaranteeing the drawn out suitability of sodium mineral usage. Developments in green mining rehearses, eco-accommodating solvents, and dependable extraction methods add to feasible sodium obtaining.

**8. Worldwide Joint efforts and Strategy Structures**

1. **Global Collaboration in Exploration**

    Tending to the difficulties and valuable open doors related with sodium mineral use requires worldwide coordinated effort. Research establishments, legislatures, and enterprises from around the world can profit from sharing information, mastery, and assets. Cooperative endeavors can speed up the advancement of feasible innovations and advance mindful sodium mineral use.

2. **Strategy Systems for Maintainable Asset The executives**

Legislatures and administrative bodies assume a urgent part in molding the supportability of sodium mineral usage. Carrying out powerful strategy structures that energize capable mining rehearses, support innovative work in feasible innovations, and guarantee the evenhanded dissemination of advantages will be instrumental in making a practical sodium industry.